I0032299

Generis

P U B L I S H I N G

GRAVITATION

- Its Cause and Mechanism -

Yoshinari Minami

Copyright © 2021 Yoshinari Minami
Copyright © 2021 Generis Publishing

All rights reserved. This book or any portion thereof may not be reproduced or used in any manner whatsoever without the written permission of the publisher except for the use of brief quotations in a book review.

Title: GRAVITATION- Its Cause and Mechanism -

ISBN : 978-1-63902-328-8

Author: Yoshinari Minami

Cover image: www.pixabay.com

Generis Publishing
Online orders: www.generis-publishing.com
Orders by email: info@generis-publishing.com

Table of Contents

PREFACE

This book describes the cause and mechanism of GRAVITATION.

From ancient Greece to modern times, philosophers and scientists have developed a history of thought processes that seek to find the cause of gravitation (gravity) that causes objects to fall. Today, Einstein explained the phenomenon of gravitation that the curved space was gravitation. General Relativity expresses the phenomenon of gravitation very strictly, but its origin and generation mechanism of the force of gravitation is not well explained; the explanation of reason is insufficient why the force of gravitation is generated when the space curves.

In this book, curved space, or curvature of space, is purely mathematical geometric quantities, but with the help of continuum mechanics, it is shown that they are related to actual forces. Given a priori assumption that space as a vacuum has a physical fine structure like continuum, it enables us to apply a continuum mechanics to the so-called "vacuum" of space.

Applying the continuum mechanics of space to General Relativity, the mechanism of gravitation has shown as one possibility.

This book is arranged in six chapters as follows.

Chapter 1 gives an overview of gravitation as an introduction.

Chapter 2 introduces what is understood about gravitation today, as well as historical considerations for gravitation. In addition, the unsolved issues of gravitation will be introduced.

Chapter 3 presents basic principles of gravitation. Here, the principle of how gravity is generated is explained as the dynamics of curved space. The principle of why apples fall to the earth is also explained using figures. We think readers can understand the reality and concept of gravitation.

Chapter 4 presents acceleration produced in curved space.

This chapter mathematically shows the principle of gravitation generation described in Chapter 3 and introduces a different derivation method of the law of universal gravitation from the conventional method and a clear principle of operation for an object falling on the earth.

Chapters 4 and 3 are the core of this book, explaining the causes and mechanisms of gravitation.

Chapter 5 presents an application to Outer Space. If we can understand the cause and mechanism of gravitation, it is expected to be applied in outer space.

Finally, Chapter 6 mentions torsion field generation by rotation, and asymmetric gravitational field. In General Relativity, since the torsion field is taken to disappear, the connection coefficients are taken to be symmetric. This chapter is provided for reference.

Although many excellent books with similar titles and similar books on aspects of gravitation or gravity have already been published, this book is a book that systematically summarizes the basic concepts and theory of gravitation or gravity.

Because even today it is not understood why gravitation is generated when space curves.

We believe this book is unique in giving readers a broad, yet in-depth view of gravitation.

Yoshinari Minami

Advanced Space Propulsion Investigation Laboratory (ASPIL)

(Formerly NEC Space Development Division), JAPAN.

1. INTRODUCTION

Gravitation is generated by curvature of space, as is well known in General Relativity. However, there is no clear mechanism of why gravitation as force is generated and acceleration is generated when the space curves. Given a priori assumption that space as a vacuum has a physical fine structure like continuum, it enables us to apply a continuum mechanics to the so-called "vacuum" of space. The pressure field derived from the geometrical structure of space is newly obtained by applying both continuum mechanics and General Relativity to space. This book is an attempt to explain the cause of gravitation and its mechanism: why acceleration arises from a curved space.

The phenomenon of falling objects has attracted people since the days of ancient Greece. Gravity, or gravitation, makes everyone interested in the essence of the phenomenon.

Why do apples fall? Why is there an attractive force between two objects? Fortunately, the formula of universal gravitational force was derived by Isaac Newton in 1665, and is phenomenologically used to explain observed facts and widely used in astronomical mechanics and spacecraft orbit calculations. However, it is still unknown why and how gravitation is generated.

The author tried to explain the cause of gravitation by applying the mechanical structure of space to General Relativity, and by applying continuum mechanics to space. The apples on the Earth will not be pulled by the Earth and fall, but will be pushed and fall in the direction of the Earth due to the pressure of the field in the curved space area around the Earth.

When we make a comparison between the space on the Earth and outer space, although there seems to be no difference, obviously a different phenomenon occurs. Simply put, an object moves radially inward, that is, drops straight down on the Earth, but in the outer space, the object floats and does not move.

The difference between the two phenomena can be explained by whether space is curved or not, that is, whether 20 independent components of a Riemann curvature tensor is zero or not. In essence, the existence of spatial curvature and curved extent region determine whether the object drops straight down or not. Although the spatial curvature at the surface of the Earth is very small value, i.e., $1.71 \times 10^{-23} (1/m^2)$, it is enough value to produce 1G (9.8 m/s^2) acceleration.

As is well known, Newton's law of universal gravitation is usually stated as that every object attracts every other object in the universe with a force that is directly proportional to the product of their masses and inversely proportional to the square of the distance between their centers.

The equation for universal gravitation thus takes the form: $F = G\dfrac{m_1 m_2}{r^2}$, where F is the gravitational force acting between two objects, m_1 and m_2 are the masses of the objects, r is the distance between the centers of their masses, and G is the gravitational constant.

This gravitational force is indirectly derived from Kepler's law based on the state of one object m_1 rotating around the other object m_2. Originally this attraction force F (i.e., gravitational force) between the two objects must be directly derived while the two objects (m_1 and m_2) are stationary.

We introduce that the nature of the curved spatial region itself creates gravitation and gravitational acceleration: that is, push an object as pressure field.

This book is a systematic compilation of the contents of the author's papers and books so far on gravity and acceleration generated in curved space regions [1-7].

The following Chapter 2 explains the History of Gravity, Chapter 3 introduces Gravitation Produced in Curved Space, Chapter 4 introduces Acceleration Produced in Curved Space, Chapter 5 presents a vision for Application to Outer Space, and Chapter 6 considers Generation of Torsion Field by Rotation as a special case.

Henceforth, the terms Gravitation and Gravity will be used in a timely manner depending on the situation. Gravitation is the attractive force between any two objects, and Gravity is the attractive force against the Earth, but they are virtually the same. There is no substantial difference.

2. HISTORY OF GRAVITY

2.1. About the Current Perception of Gravity

From ancient Greece to modern times, philosophers and scientists have developed a history of thought processes that seek to find the cause of gravity that causes objects to fall. After all, today's theory of gravity was established by Copernicus, Kepler, Galilei, Newton, and Einstein. Einstein explained the phenomenon of gravity that the curved space was gravity. Although the explanation is correct, the interpretation of "the force of gravity" that Copernicus, Kepler, Galilei, Newton and others emphasized from the days of ancient Greece was lost. He compromises in the easy concept of "resistance force of space-time", that is, gravity is only felt when you resist space-time.

In General Relativity, the phenomenology of gravity is well explained, but the origin of gravity is only the curved space, and there is no explanation as to why things move and fall when space curves, and detailed examination has not been done. There is still room for future development. It is our experimental rule that the action of force is necessary for an object to move or start moving. This is a pending problem in General Relativity. That is, there is no explanation as to why force is generated when space curves.

According to Einstein's theory of General Relativity, the following Fig.1 is often seen and explained in general books, but it is an unfriendly explanation. After all, it can only be understood that a small object is drawn into the space-time depression around the Earth and falls, and lacks a clear principle of why the small object falls.

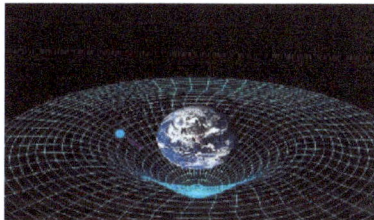

Fig. 1. Curved space around the Earth

(https://ord.yahoo.co.jp/o/image/)

Concerning the explanation of Fig.1, an explanatory drawing depicts the state where the Earth is dropped into a two-dimensional grid pattern plane. As can be seen from the explanatory drawing, a state in which the lattice pattern is distorted can be visually recognized, and the distorted (or curved) lattice pattern itself can be interpreted as gravity. If this illustration is likened to ordinary, it is the same as a heavy object sinking on a trampoline.

In this figure and description, the flat grid-like space is depressed due to the mass of the Earth. It's just a misconception that the small object rolls toward the periphery of the dent, but it is not a consistent explanation of the relationship between the Earth, the curved space, and the falling small object.

Perhaps this figure seems to be stated where the plane of the lattice pattern drawn in two dimensions is dented and the lattice pattern is distorted, that is, the distorted lattice, like the state where the Earth sinks on the trampoline. The pattern itself is gravity. After all, it's not a convincing explanation, because we have the impression that small objects are drawn into the hollow space around the Earth and fall down.

In General Relativity, geodesic is a generalization of a "straight line" over a curved space-time. The world line of particles that do not receive any external force other than gravity is a kind of geodesic line and is important. In other words, particles in free motion or free fall move along the geodesic line. However, in general, movement requires a force to move. There is no mechanism to explain the force generation.

The following section explains "History of Trying to Find the Cause of Gravity".

2.2. History of Trying to Find the Cause of Gravity

Let's take a brief look at the process of consideration by philosophers and scientists about the phenomenon of gravity that causes all objects to fall. In the Greek era, they were interested in why objects fall, as well as considering the view of the universe. Aristarchos's view of the universe first showed the idea of the Copernican theory that the Sun is at the center of the universe, and the Earth and planets orbit around it in the same way.

Although we think that the ideas of Aristarchos from 2300 years ago are natural for us today, they were ignored at that time and criticized by Aristoteles. Aristoteles established an earth-centered cosmology that all objects have weight and fall toward the Earth. It is a rationale for the Earth being the center of the universe. Aristoteles sought the reason for the fall phenomenon from the root element of matter, and established cosmology that the Earth is the center of the universe.

Later, Copernicus advocated sun-centered cosmology and the Copernican theory appeared. It is the reappearance of the Copernican theory of Aristarchos. However, it has a drawback that it is much less accurate than the Earth-centered space model of Ptolemaeos that mathematically completed the mainstream theory.

Kepler discovered the three laws of Kepler's planetary motion based on the vast amount of precision astronomical observation data accumulated by astronomer Brahe. From this point, the refinement of the Copernican theory by correct orbit calculation begins.

From the experiment of the falling motion of rolling an iron ball on the slope by Galilei, the relationship between speed, acceleration and time was shown and the theory of inertia was established.

Batons for elucidating the phenomenon of gravity were handed to Copernicus, Kepler, Galilei, and Newton in that order. By comparing the falling apple and the moon around the Earth, and thinking that the moon is also falling on the Earth, Newton selects Kepler's theory that attractive force acts across the space. The three laws of Kepler's planetary motion were derived from observational data, but it was unclear why those laws hold. However, the universal gravitational formula by Newton brought about three laws of Kepler's planetary motion mathematically. This one law of universal gravitation explained the falling phenomenon and the motion of the celestial body at the same time. The law of universal gravitation was positioned as a definite law explaining the force of gravity. And the new baton was handed over to Einstein from Newton.

2.3. Remaining Problem in General Relativity

General Relativity explains the phenomenon of gravity, but at present it does not explain why gravity occurs. Unfortunately, General Relativity seems to have compromised the explanation of the phenomenon of force.

As mentioned before, geodesic is a generalization of a "straight line" over a curved space-time. The world line of particles that do not receive any external force other than gravity is a kind of geodesic line and is important. In other words, particles in free motion or free fall move along the geodesic line. However, in general, the movement requires force to move or to start moving. There is no mechanism to explain this force generation (Fig. 2 (a)).

Furthermore, in General Relativity, gravity is thought to be a consequence of the geometry of the curved space-time, not of force, and the source of the space-time curvature is the energy momentum tensor (for example, representing matter). Thus, the orbit of a planet orbiting a star is a projection of a geodesic on a curved four-dimensional space-time into three-dimensional space (Fig. 2 (b)).

Where did the force go? Where did the force exerted by gravity, which plagued many philosophers and scientists such as "gravity is a force", gone?

Einstein came to think of gravity as a kind of semantic illusion that can only be felt when resisting space-time. What is the relationship between curved space-time and gravity? Gravity was the resistance that appeared because the hand that was supporting the object trying to flow as it was in the space-time interfered. If you leave yourself to something like a flowing stream of curved space-time, you will not feel the force as resistance (just like as free fall). A force as gravity arises only when countering a flow.

This explanation seems to be an unconvincing compromise for the gravity.

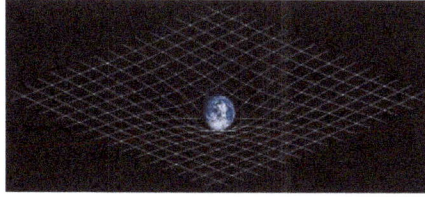

(a) A hollow space around the Earth

(https://ja.wikipedia.org/wiki/)

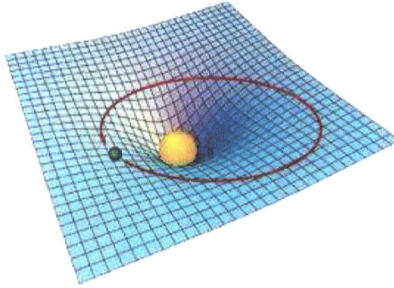

(b) An object that falls to the bottom of a hollow space

(https://ord.yahoo.co.jp/o/image/)

Fig. 2. Explanation of gravity by General Relativity

General Relativity expresses the phenomenon of gravity very strictly and correctly, but the origin and generation mechanism of the force of gravity are insufficient. The current situation is that there is no convincing explanation of why gravity occurs when the space curves.

It is considered that there is room for future development as the remaining issues to be considered.

3. GRAVITATION PRODUCED IN CURVED SPACE

In this chapter, we will explain the force generated by curved space and the mechanism of gravitation, i.e., basic principles of gravitation. Here, the principle of how gravity is generated is explained as the dynamics of curved space. The principle of why apples fall to the Earth is also explained using figures.

3.1. Generation of Surface Force Induced by Curved Space

On the supposition that space is an infinite continuum, continuum mechanics can be applied to the so-called "vacuum" of space. This means that space can be considered as a kind of transparent field with elastic properties. Consider a thin layer of a single space obtained by slicing the space of a transparent rubber block.

If space curves, then an inward normal stress "$-P$" is generated. This normal stress, i.e., surface force serves as a sort of pressure field as shown in Fig.3.1.

$$-P = N \cdot (2R^{00})^{1/2} = N \cdot (1/R_1 + 1/R_2), \tag{3.1}$$

where N is the line stress, R_1, R_2 are the radius of principal curvature of curved surface, and R^{00} is the major component of spatial curvature.

A large number of curved thin layers form the unidirectional surface force as shown in Fig.3.2.

When surface forces are accumulated, a surface force field, that is, a force field is created. An object in the force field is accelerated by the force, so an acceleration field is generated. A large number of curved thin layers form the unidirectional surface force, i.e., acceleration field. Accordingly, the spatial curvature R^{00} produces the acceleration field α.

It is now understood that the membrane force on the curved surface and each principal curvature generates the normal stress "$-P$" with its direction normal to the curved surface as a surface force. The normal stress "$-P$" acts towards the inside of the surface as shown in Fig.3.1.

A thin-layer of curved surface will take into consideration within a spherical space having a radius of R and the principal radii of curvature that are equal to the radius ($R_1=R_2=R$). Since the membrane force N (serving as the line stress) can be assumed to have a constant value, Eq. (3.1) indicates that the curvature R^{00} generates the inward normal stress P of the curved surface. The inwardly directed normal stress serves as a pressure field.

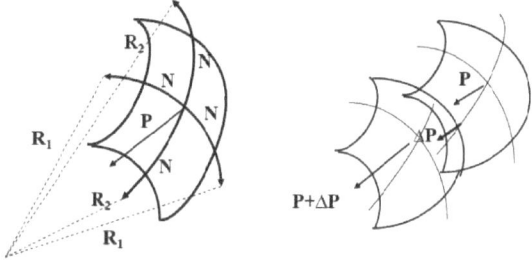

Fig. 3.1. Curvature of space plays a significant role. If space curves, then inward stress (surface force) "P" is generated. A sort of pressure field.

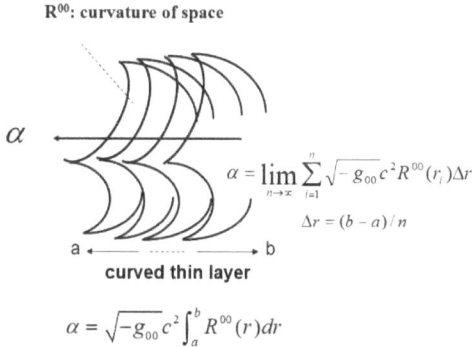

$$\alpha = \lim_{n \to \infty} \sum_{i=1}^{n} \sqrt{-g_{00}} \, c^2 R^{00}(r_i) \Delta r$$

$$\Delta r = (b-a)/n$$

curved thin layer

$$\alpha = \sqrt{-g_{00}} \, c^2 \int_a^b R^{00}(r)dr$$

Fig. 3.2. A large number of curved thin layers form the unidirectional surface force, i.e., acceleration field α .

When the curved surfaces are included in a great number, some type of unidirectional pressure field is formed. A region of curved space is made of a large number of curved surfaces and they form the field as a unidirectional surface force (i.e. normal stress). Since the field of the surface force is the field of a kind of force, the force accelerates matter in the field, i.e., we can regard the field of the surface force as the acceleration field. A large number of curved thin layers form the unidirectional acceleration field (Fig.3.2).

Accordingly, the spatial curvature R^{00} produces the pressure field (i.e., acceleration field α).

For example, consider a soap bubble (Fig.3.3).

The pressure "P" due to the membrane force N on the surface of a soap bubble of radius R is directed inward. The membrane force on the surface of the soap bubble corresponds to N in the Fig.3.1.

$$-P = N \cdot (1/R_1 + 1/R_2) = N \cdot (1/R + 1/R) = 2N/R. \tag{3.2}$$

This pressure "P" keeps the soap bubbles from breaking due to the expansion force of the internal air.

Next considering the dynamics of the surface of a soap bubble, we can see the similarity of gravity generation. Fig.3.1 shows the basic concept of the gravity generation mechanism. We show that the curvature of space creates a pressure field as an acceleration field. As explained before, Fig.3.1 shows that the vertical force P (surface force) toward the center of the surface is generated by the membrane force (line stress) N and the radii of curvature R_1 and R_2 of the thin layer with a curved space. The radius of curvature decreases toward the inner side, and the vertical stress P (surface force) increases. The surface force of the membrane layer becomes the membrane pressure.

Fig.3.3 shows the surface force P toward the center of the soap bubble. The surface of the soap bubble extends due to surface tension (line stress N) to maintain the shape of the soap bubble, but it is known from continuum mechanics that the surface force P toward the center of the soap bubble is working. If this is applied to the space as it is, the space curved in a spherical shape applies pressure toward the center of the sphere. This is the reality of gravity.

Fig.3.2 shows that when a large number of curved thin film layers are integrated, unidirectional surface forces are integrated to form a pressure field, i.e., an acceleration field. It can be used to calculate the gravitational acceleration.

Fig.3.4 shows how the surface force due to the accumulation of many curved surfaces pushes the apple.

Fig. 3.3. Surface force toward the center of the soap bubble

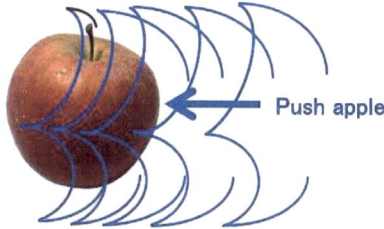

Fig. 3.4. Surface forces of multiple curved thin film layers push apple

The fundamental three-dimensional space structure is determined by quadratic surface structure. Therefore, Gaussian curvature K in two-dimensional Riemann space is significant. The relationship between Gaussian curvature K and the major component of spatial curvature R^{00} is given by:

$$K = \frac{R_{1212}}{(g_{11}g_{22} - g_{12}{}^2)} = \frac{1}{2} \cdot R^{00}, \tag{3.3}$$

where R_{1212} is non-zero component of Riemann curvature tensor.

Applying membrane theory, the following equilibrium conditions are obtained in quadratic surface, given by:

$$N^{\alpha\beta} b_{\alpha\beta} + P = 0, \tag{3.4}$$

where $N^{\alpha\beta}$ is a membrane force, i.e., line stress of curved space, $b_{\alpha\beta}$ is second fundamental metric of curved surface, and P is the normal stress on curved surface.

The second fundamental metric of curved space $b_{\alpha\beta}$ and principal curvature $K_{(i)}$ has the following relationship using the metric tensor $g_{\alpha\beta}$,

$$b_{\alpha\beta} = K_{(i)} g_{\alpha\beta}. \tag{3.5}$$

Therefore we get:

$$N^{\alpha\beta} b_{\alpha\beta} = N^{\alpha\beta} K_{(i)} g_{\alpha\beta} = g_{\alpha\beta} N^{\alpha\beta} K_{(i)} = N_\alpha{}^\alpha K_{(i)} = N \cdot K_{(i)}. \tag{3.6}$$

From Eq. (3.4) and Eq. (3.6), we get:

$$N_\alpha{}^\alpha K_{(i)} = -P. \tag{3.7}$$

As for the quadratic surface, the indices α and i take two different values, i.e., 1 and 2, therefore Eq. (3.7) becomes:

$$N_1{}^1 K_{(1)} + N_2{}^2 K_{(2)} = -P, \tag{3.8}$$

where $K_{(1)}$ and $K_{(2)}$ are principal curvature of curved surface and are inverse number of radius of principal curvature (i.e. $1/R_1$ and $1/R_2$).

The Gaussian curvature K is represented as:

$$K = K_{(1)} \cdot K_{(2)} = (1/R_1) \cdot (1/R_2). \tag{3.9}$$

Accordingly, suppose $N_1{}^1 = N_2{}^2 = N$, we get:

$$N \cdot (1/R_1 + 1/R_2) = -P. \tag{3.10}$$

It is now understood that the membrane force on the curved surface and each principal curvature generate the normal stress "$-P$" with its direction normal to the curved surface as a surface force. The normal stress $-P$ is towards the inside of surface as showing in Fig.3.1.

A thin-layer of curved surface will be taken into consideration within a spherical space having a radius of R and the principal radii of curvature which are equal to the radius $(R_1=R_2=R)$. From Eqs. (3.3) and (3.9), we then get:

$$K = \frac{1}{R_1} \cdot \frac{1}{R_2} = \frac{1}{R^2} = \frac{R^{00}}{2}. \tag{3.11}$$

Considering $N \cdot (2/R) = -P$ of Eq. (3.10), and substituting Eq. (3.11) into Eq. (3.10), the following equation is obtained:

$$-P = N \cdot \sqrt{2R^{00}}. \tag{3.12}$$

Since the membrane force N (serving as the line stress) can be assumed to have a constant value, Eq. (3.12) indicates that the curvature R^{00} generates the inward normal stress P of the curved surface. The inwardly directed normal stress serves as a kind of pressure field. When the curved surfaces are included in great number, some type of unidirectional pressure field is formed. A region of curved space is made of a large number of curved surfaces and they form the field of unidirectional surface force (i.e. normal stress). Since the field of surface force is the field of a kind of force, a body in the field is accelerated by the force, i.e., we can regard the field of surface force as the acceleration field. Accordingly, the cumulated curved region of curvature R^{00} produces the acceleration field α.

Here, we give an account of curvature R^{00} in advance. The solution of metric tensor $g^{\mu\nu}$ is found by gravitational field equation as the following:

$$R^{\mu\nu} - \frac{1}{2} \cdot g^{\mu\nu} R = -\frac{8\pi G}{c^4} \cdot T^{\mu\nu}, \tag{3.13}$$

where $R^{\mu\nu}$ is the Ricci tensor, R is the scalar curvature, G is the gravitational constant, c is the speed of light, $T^{\mu\nu}$ is the energy momentum tensor. Furthermore, we have the following relation for scalar curvature R:

$$R = R^{\alpha}{}_{\alpha} = g^{\alpha\beta} R_{\alpha\beta}, \ R^{\mu\nu} = g^{\mu\alpha} g^{\nu\beta} R_{\alpha\beta}, \ R_{\alpha\beta} = R^{i}{}_{\alpha i \beta} = g^{ij} R_{i\alpha j\beta}. \tag{3.14}$$

Ricci tensor $R^{\mu\nu}$ is represented by:

$$R_{\mu\nu} = \Gamma^{\alpha}{}_{\mu\alpha,\nu} - \Gamma^{\alpha}{}_{\mu\nu,\alpha} - \Gamma^{\alpha}{}_{\mu\nu}\Gamma^{\beta}{}_{\alpha\beta} + \Gamma^{\alpha}{}_{\mu\beta}\Gamma^{\beta}{}_{\nu\alpha} \quad (=R_{\nu\mu}), \tag{3.15}$$

where Γ^i_{jk} is Riemannian connection coefficient.

If the curvature of space is very small, the term of higher order than the second can be neglected, and Ricci tensor becomes:

$$R_{\mu\nu} = \Gamma^\alpha_{\mu\alpha,\nu} - \Gamma^\alpha_{\mu\nu,\alpha}. \tag{3.16}$$

The major curvature of Ricci tensor ($\mu = \nu = 0$) is calculated as follows:

$$R^{00} = g^{00}g^{00}R_{00} = -1 \times -1 \times R_{00} = R_{00}. \tag{3.17}$$

As previously mentioned, Riemannian geometry is a geometry that deals with a curved Riemann space, therefore Riemann curvature tensor is the principal quantity. All components of Riemann curvature tensor are zero for flat space and non-zero for curved space. If an only non-zero component of Riemann curvature tensor exists, the space is not flat space but curved space. Therefore, the curvature of space plays a significant role.

<Supplemental explanation for Eq. (3.3)>

For a two-dimensional surface, from the Bianchi identity, the Riemann curvature tensor is given by:

$$R_{\nu\mu\lambda\kappa} = K(g_{\nu\lambda}g_{\mu\kappa} - g_{\nu\kappa}g_{\mu\lambda}),$$

that is, $R_{1212} = K(g_{11}g_{22} - g_{12}^2)$.

And, for a spherical surface of radius r, its Gaussian curvature K is $1/r^2$. The scalar curvature R and the Gaussian curvature K on the quadratic surface are as follows:

$$R = R_i{}^i = g^{ij}R_{ij} = g^{11}R_{11} + g^{12}R_{12} + g^{21}R_{21} + g^{22}R_{22}$$
$$= \frac{1}{r^2}(-1) + \frac{1}{r^2\sin^2\theta}(-\sin^2\theta) = -\frac{2}{r^2} = -2K.$$

Calculating metric and Riemannian connection coefficient in spherical coordinate system, and using $R_{\mu\nu} = R^\sigma_{\mu\sigma\nu} = g^{\rho\sigma}R_{\rho\mu\sigma\nu}$, then calculate the Ricci tensor, the following are obtained:

$$g^{11} = \frac{1}{r^2}, \quad g^{22} = \frac{1}{r^2 \sin^2 \theta}, \quad \text{other } g^{\mu\nu} = 0.$$

Using $R_{\rho\sigma\mu\nu} = g_{\rho\lambda}R^{\lambda}_{\ \sigma\mu\nu}$, there is only one independent component

$$R_{1212} = -r^2 \sin^2 \theta.$$

Namely, from $R_{\mu\nu} = g^{\rho\sigma} R_{\rho\mu\sigma\nu}$, $R_{11} = g^{22}R_{2121} = g^{22}R_{1212} = \dfrac{-r^2 \sin^2 \theta}{r^2 \sin^2 \theta} = -1$,

$R_{12} = g^{11}R_{1112} = 0, R_{21} = g^{11}R_{1211} = 0$, $R_{22} = g^{11}R_{1212} = \dfrac{1}{r^2}(-r^2 \sin^2 \theta) = -\sin^2 \theta$.

Thus $R_{11} = -1$, $R_{22} = -\sin^2 \theta$, $R_{12} = R_{21} = 0$.

On the while, the scalar curvature R $(1/m^2)$ on a four-dimensional surface is given by:

$$R = R^{i}_{\ i} = g_{ij}R^{ij} = g_{00}R^{00} + g_{11}R^{11} + g_{22}R^{22} + g_{33}R^{33}$$
$$\approx g_{00}R^{00} = -R^{00} \ (g_{00} \approx -1: \text{ weak field})$$

Accordingly, $R = -2K = -R^{00}$, then $K = \dfrac{1}{2} \cdot R^{00}$ is obtained.

3.2. Mechanism of GRAVITATION: as a Pressure Field Induced by Curved Space

As shown in Fig.3.5, the gravitational field around the Earth is multiply covered by concentric or spherical curved spaces centered on the Earth. Considering the case of the Earth, the curvature of space is spherically symmetric about the Earth and is fixed to the Earth, so the Earth itself cannot move due to the curvature of the space generated by the Earth.

However, as shown in Fig.3.6, the apples on the Earth are independently in the curved spatial region of the Earth. Since the apple exists in the curved spatial region from the curved spatial layer at the apple's position to the curved spatial layer at the distant position, the apple is pushed by the generated curved space (i.e., pressure) and falls. That is, referring to Fig.3.7, a sort of graduated pressure field is generated by the curved range from an arbitrary point "a" in curved space to a point "b" (the point at which space is absent of curvature, i.e., flat space of curvature "0"). Then apple moves directly towards the center of the Earth, that is, the apple falls. Falling acceleration of apple in curved space is proportional to both the value of spatial curvature and the size of curved space.

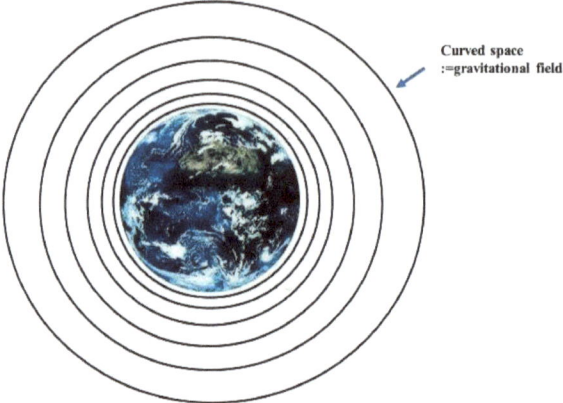

Fig. 3.5. Gravitational field around the Earth is a curved space that is concentric or spherical about the Earth

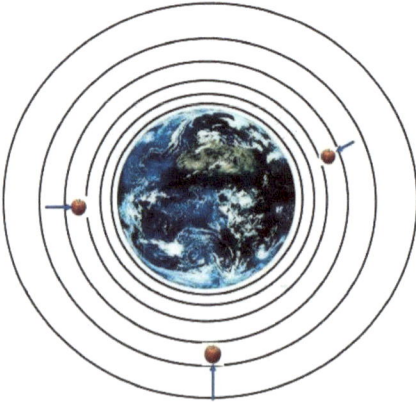

Fig. 3.6. Since the apples on the Earth are independently in the curved spatial regions of the Earth, the apples fall under the pressure generated in the curved spatial regions

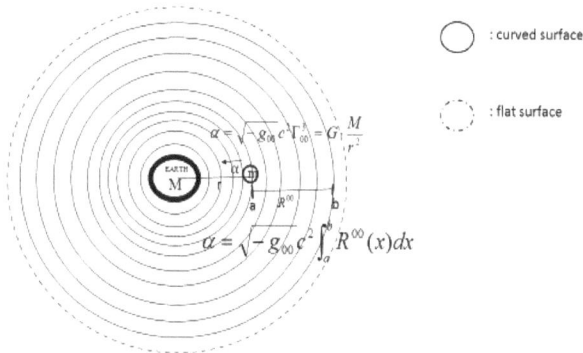

Fig. 3.7. Apple falls receiving a pressure of the field

Next, consider the universal gravitational force of a well-known apple falling to the Earth.

Although the attraction between the Earth and the apple by universal gravitation can be explained by a mathematical formula, $F = G\dfrac{Mm}{r^2}$, there is no explanation of the mechanism of the attraction, that is, the principle of operation.

The mechanism can be understood by interpreting that the Earth and the apple are pushed toward each other from behind the curved space area around the Earth and the curved space area around the apple.

This phenomenon is that an apple is not pulled and falls by the Earth, but the apple is pushed toward the Earth under the pressure of the vast curved space area of the Earth.

Fig.3.8 shows the mechanism.

In the upper diagram of Fig.3.8, there are mass bodies A and B, and the space around each mass body is curved. As already explained, the mass B is pushed out of the curved space field generated by the mass A, and the mass A is pushed out of the curved space field generated by the mass B, so that they will move in the direction of opposition to each other. In the lower diagram of Fig.3.8, mass A is a giant mass of the Earth, and mass B is a light apple.

Apple is pushed from the vast curved space area of the Earth and go straight to the Earth. On the other hand, the Earth is also pushed from the narrow-curved space area of the apple and go straight to the apple. Since the

19

mass of an apple is smaller than that of the Earth, the range of the curved space is small and the acceleration with respect to the Earth is almost zero.

In effect, it looks like an apple is pulled by the Earth and falls. Please refer to Ref. [4] in detail.

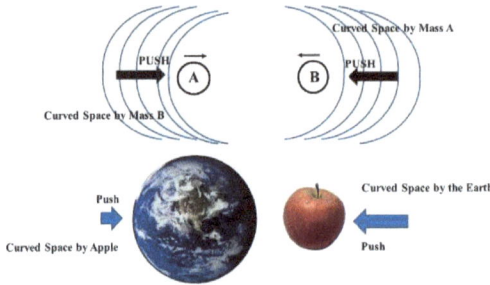

Fig. 3.8. Apple and the Earth are pushed out of a curved space and collide

3.3. Consideration of Gravitation

Let us consider about gravitation. Why does apple fall in the Earth? A well-known answer is that there exists gravitation between Earth and apple. Apple is because it's pulled by a law of universal gravitation $F = G\dfrac{Mm}{r^2}$ to the Earth. Here, M is the mass of Earth, m is the mass of apple, G is the gravitational constant, r is the distance between Earth and apple, F is the gravitational force. From a phenomenological standpoint, it is a sufficient explanation.

However, what is the mechanism? According to General Relativity, it is said that apple moves geodesic line formed by curved space near the Earth. This is seen as lacking in sufficient explanation. The following explanation may allow someone to understand the mechanism of gravitation.

If we were to visualize the curvature of space around the Earth (M), we would describe it as having an aggregation of curved surface. A great number of thin curved surfaces are arranged in a spherical concentric pattern. This curvature would gradually become smaller as we moved away from the Earth in what we could imagine as layers of an onion. The surrounding space becomes a flat space of curvature "0" at an imagined immense distance from the Earth (Fig.3.9).

In the following thought experiment, an apple of mass "m" positioned at a distance r apart from the Earth would receive a pressure of the field formed by an accumulation of the normal stress (Fig.3.2). As was described earlier, with reference to Fig.3.1, the membrane force on the curved surface and each principal curvature generates the normal stress "–P" with its direction normal to the curved surface as a surface force. The normal stress "–P" acts towards the inside of the surface as shown in Fig.3.1. Here the negative sign only indicates the direction. Think of "P" heading towards the center.

A thin-layer of curved surface will take into consideration within a spherical space having a radius of R and the principal radii of curvature that are equal to the radius ($R_1=R_2=R$). Since the membrane force N (serving as the line stress) can be assumed to have a constant value, the inwardly directed normal stress serves as a pressure field. When the curved surfaces are included in a great number, some type of unidirectional pressure field is formed.

That is, a sort of graduated pressure field is generated by the curved range from an imaginary point "a" in curved space to a point "b" (the point at which space is absent of curvature, i.e., flat space of curvature "0") (Fig.3.9). Then apple moves directly towards the center of the Earth, that is, the apple falls. Falling acceleration of apple in curved space is proportional to both the value of spatial curvature and the size of curved space.

If the Earth (M) were to disappear instantly, the curved surface of space close to the Earth would return to the flat surface. Because an external action causing curvature (i.e., mass energy) disappears. The change from a curved surface to a flat surface would advance for the position r of the apple at the speed of light (i.e., the strain rate of space-time). The propagation velocity of the change from flat space to curved space and the propagation velocity of change from curved space to flat space are both the same, i.e., the velocity of light.

However, in our thought experiment, the apple would still receive pressure from the surrounding field by the accumulation of the normal stress. Because, since there still exists the curved region behind the apple from a to b (the remote flat space), the apple continues falling. The pressure continues to push the apple to the center of the Earth (Fig.3.10).

As soon as the change from a curved surface to a flat surface passes through the point of the apple (i.e., "a" point), the pressure at point "a" disappears and the apple would only float without falling (Fig.3.11). Because apples no longer exist in curved spatial areas.

The above discussion provides a basis to consider the following thought experiment. Even if the Sun instantly disappeared, the Earth would still continue to revolve around the Sun until 8 minutes 32 seconds, or the time at which it takes light to advance between the Sun and the Earth. However, as soon as the change from curved surface to flat surface passes through the point of the Earth, or at 8 minutes 32 seconds after the event, the pressure pushing the Earth would disappear, and the Earth would fly away in the tangential direction of its orbit.

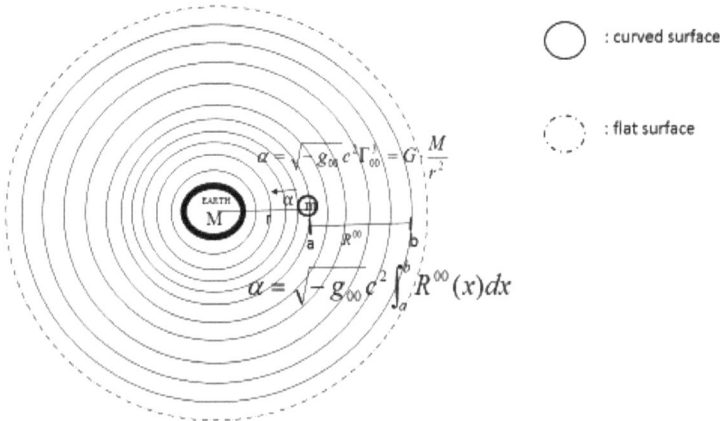

$$\alpha = \sqrt{-g_{00}}\, c^4 \Gamma^h_{00} = G_1 \frac{M}{r^2}$$

$$\alpha = \sqrt{-g_{00}}\, c^2 \int_a^b R^{00}(x)\,dx$$

○ : curved surface

⬭ : flat surface

Fig. 3.9. Apple falls receiving a pressure of the field

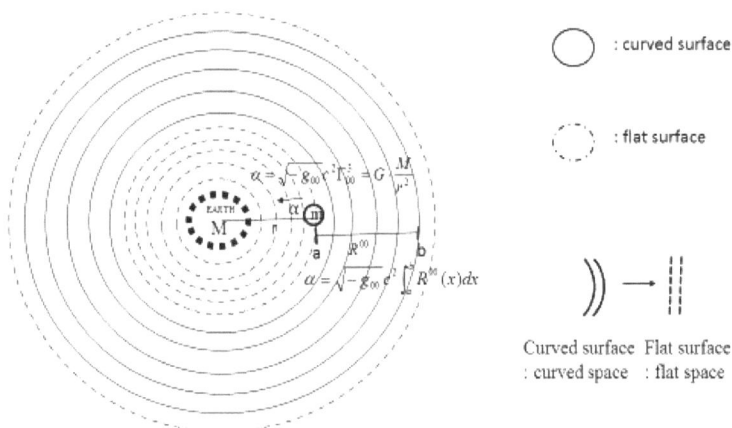

$\alpha = \sqrt{-g_{00}}\, c^2\, \Gamma_{00}^i = G\dfrac{M}{r^2}$

$\alpha = \sqrt{-g_{00}}\, c^2\, \displaystyle\int_a^b R^{00}(x)dx$

○ : curved surface

◌ : flat surface

)) → ||

Curved surface Flat surface
: curved space : flat space

Fig. 3.10. Apple still continues falling receiving a pressure of the field

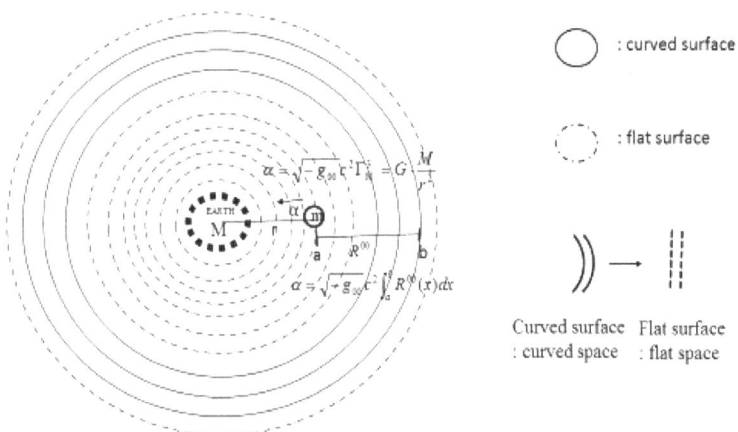

$\alpha = \sqrt{-g_{00}}\, c^2\, \Gamma_{00}^i = G\dfrac{M}{r^2}$

$\alpha = \sqrt{-g_{00}}\, c^2\, \displaystyle\int_a^b R^{00}(x)dx$

○ : curved surface

◌ : flat surface

)) → ||

Curved surface Flat surface
: curved space : flat space

Fig. 3.11. Apple only floats without falling due to lack of pressure of the field

In view of this, gravitation may be considered as a pressure generated in a region of curved space.

4. ACCELERATION PRODUCED IN CURVED SPACE

This chapter presents the acceleration produced in curved space. We mathematically show the principle of gravitation generation described in Chapter 3 and introduces a different derivation method of the law of universal gravitation from the conventional method and a clear principle of operation for an object falling on the Earth.

4.1. Derivation of Acceleration

A massive body causes the curvature of space-time around it, and a free particle responds by moving along a geodesic line in that space-time. The path of free particle is a geodesic line in space-time and is given by the following geodesic equation:

$$\frac{d^2x^i}{d\tau^2} + \Gamma^i_{jk} \cdot \frac{dx^j}{d\tau} \cdot \frac{dx^k}{d\tau} = 0, \tag{4.1}$$

where Γ^i_{jk} is Riemannian connection coefficient, τ is proper time, x^i is four-dimensional Riemann space, that is, three dimensional space ($x=x^1$, $y=x^2$, $z=x^3$) and one dimensional time ($w=ct=x^0$), c is the velocity of light. These four coordinate axes are denoted as x^i ($i=0, 1, 2, 3$).

Proper time is the time to be measured in a clock resting for a coordinate system. We have the following relation derived from an invariant line element ds^2 between Special Relativity (flat space) and General Relativity (curved space):

$$d\tau = \sqrt{-g_{00}}dx^0 = \sqrt{-g_{00}}cdt. \tag{4.2}$$

From Eq. (4.1), the acceleration of free particle is obtained by:

$$\alpha^i = \frac{d^2x^i}{d\tau^2} = -\Gamma^i_{jk} \cdot \frac{dx^j}{d\tau} \cdot \frac{dx^k}{d\tau}. \tag{4.3}$$

As is well known in General Relativity, in the curved space region, the massive body "m (kg)" existing in the acceleration field is subjected to the following force F^i (N):

$$F^i = m\Gamma^i_{jk} \cdot \frac{dx^j}{d\tau} \cdot \frac{dx^k}{d\tau} = m\sqrt{-g_{00}}c^2\Gamma^i_{jk}u^ju^k = m\alpha^i, \tag{4.4}$$

where u^j, u^k are the four velocity, Γ^i_{jk} is the Riemannian connection coefficient, and τ is the proper time.

From Eqs. (4.3), (4.4), we obtain:

$$\alpha^i = \frac{d^2 x^i}{d\tau^2} = -\Gamma^i_{jk} \cdot \frac{dx^j}{d\tau} \cdot \frac{dx^k}{d\tau} = -\sqrt{-g_{00}} c^2 \Gamma^i_{jk} u^j u^k . \tag{4.5}$$

Eq. (4.5) yields a more simple equation from the condition of linear approximation, that is, weak-field, quasi-static, and slow motion (speed v $\ll$ speed of light c: $u^0 \approx 1$):

$$\alpha^i = -\sqrt{-g_{00}} \cdot c^2 \Gamma^i_{00} . \tag{4.6}$$

On the other hand, the major component of spatial curvature R^{00} in the weak field is given by:

$$R^{00} \approx R_{00} = R^\mu_{0\mu 0} = \partial_0 \Gamma^\mu_{0\mu} - \partial_\mu \Gamma^\mu_{00} + \Gamma^\nu_{0\mu} \Gamma^\mu_{\nu 0} - \Gamma^\nu_{00} \Gamma^\mu_{\nu\mu} . \tag{4.7}$$

In the nearly Cartesian coordinate system, the value of $\Gamma^\mu_{\nu\rho}$ are small, so we can neglect the last two terms in Eq. (4.7), and using the quasi-static condition we get:

$$R^{00} = -\partial_\mu \Gamma^\mu_{00} = -\partial_i \Gamma^i_{00} . \tag{4.8}$$

From Eq. (4.8), we get formally:

$$\Gamma^i_{00} = -\int R^{00}(x^i) dx^i . \tag{4.9}$$

Substituting Eq. (4.9) into Eq. (4.6), we obtain:

$$\alpha^i = \sqrt{-g_{00}} c^2 \int R^{00}(x^i) dx^i . \tag{4.10}$$

Accordingly, from the following linear approximation scheme for the gravitational field equation:(1) weak gravitational field, i.e., small curvature limit, (2) quasi-static, (3) slow-motion approximation (i.e., $v/c \ll 1$), and considering range of curved region, we get the following relation between acceleration of curved space and curvature of space:

$$\alpha^i = \sqrt{-g_{00}} c^2 \int_a^b R^{00}(x^i) dx^i , \tag{4.11}$$

where α^i: acceleration (m/s²), g_{00}: time component of metric tensor, a-b: range of curved space (m), x^i: components of coordinate ($i=0,1,2,3$), c: velocity of light, R^{00}: major component of spatial curvature($1/m^2$).

Eq. (4.11) indicates that the acceleration field α^i is produced in curved space. The intensity of acceleration produced in curved space is proportional to the product of spatial curvature R^{00} and the length of curved region [from a to b].

Eq. (4.4) yields more simple and effective equation from above-stated linear approximation ($u^0 \approx 1$),

$$F^i = m\sqrt{-g_{00}}\,c^2\Gamma^i_{00}u^0u^0 = m\sqrt{-g_{00}}\,c^2\Gamma^i_{00} = m\alpha^i$$
$$= m\sqrt{-g_{00}}\,c^2\int_a^b R^{00}(x^i)dx^i$$

(4.12)

Setting $i=3$ (i.e., direction of radius of curvature: r), we get Newton's second law:

$$F^3 = F = m\alpha = m\sqrt{-g_{00}}\,c^2\int_a^b R^{00}(r)dr = m\sqrt{-g_{00}}\,c^2\Gamma^3_{00}.$$

(4.13)

The acceleration (α) of curved space and its Riemannian connection coefficient (Γ^3_{00}) are given by:

$$\alpha = \sqrt{-g_{00}}\,c^2\Gamma^3_{00}, \quad \Gamma^3_{00} = \frac{-g_{00,3}}{2g_{33}}.$$

(4.14)

where c: velocity of light, g_{00} and g_{33}: component of metric tensor,

$g_{00,3} : \partial g_{00}/\partial x^3 = \partial g_{00}/\partial r$. We choose the spherical coordinates "$ct=x^0$, $r=x^3$, $\theta=x^1$, $\varphi=x^2$" in space-time.

The acceleration α is represented by the equation both in the differential form and in the integral form. Practically, since the metric is usually given by the solution of gravitational field equation, the differential form has been found to be advantageous.

<Supplemental explanation for Eq. (4.2, 4.5)>

For the flat space (Special Relativity), invariant line element ds^2 is described in

$$ds^2 = \eta_{\mu\nu}dx^\mu dx^\nu .$$

In the case of dx^μ is time-like,

$$ds^2 = \eta_{00}dx^0 dx^0 = -(dx^0)^2 = -(d\tau)^2 .$$

Here, $\eta_{00} = -1$ $dx^0 = d\tau = cdt$.

$d\tau$ is a proper time measured by a clock that is stationary with respect to the coordinate system.

The proper time does not depend on the coordinate system.

For the curved space (General Relativity), invariant line element ds^2 is described in

$$ds^2 = g_{\mu\nu}(x)dx^\mu dx^\nu .$$

In the case of dx^μ is time-like,

$$ds^2 = g_{00}dx^0 dx^0 = g_{00}(dx^0)^2 .$$

We have the following relation derived from an invariant line element ds^2 between Special Relativity (flat space) and General Relativity (curved space): since the infinitesimal line element ds^2 is invariant,

$$-(d\tau)^2 = g_{00}(dx^0)^2 ,$$

then using one dimensional time (w=ct=x⁰), we get:

$$d\tau = \sqrt{-g_{00}}\,dx^0 = \sqrt{-g_{0u}}\,dw = \sqrt{-g_{00}}\,cdt .$$

By performing the second derivate,

$$d^2\tau = \sqrt{-g_{00}}\,(dx^0)^2 = \sqrt{-g_{00}}\,dw^2 = \sqrt{-g_{00}}\,c^2 dt^2 .$$

Further, as is known well in Special Relativity, in the case of slow-motion approximation (i.e., $v/c \ll 1$), four velocity $u^\mu = \dfrac{dx^\mu}{d\tau}$ becomes the following:

$$u^i = \frac{1}{c}\gamma v^i \approx 0, u^0 = \gamma = \sqrt{1 - (v/c)^2} \approx 1(v \ll c).$$

Here, substitute the above equation into Eq. (4.3),

$$\alpha^i = \frac{d^2 x^i}{d\tau^2} = -\Gamma^i_{jk} \cdot \frac{dx^j}{d\tau} \cdot \frac{dx^k}{d\tau} = -\Gamma^i_{jk} u^j u^k .$$

Namely, $\dfrac{d^2 x^i}{d\tau^2} = \dfrac{d^2 x^i}{\sqrt{-g_{00}}c^2 dt^2} = -\Gamma^i_{jk} u^j u^k$, then we get, $\dfrac{d^2 x^i}{dt^2} = -\sqrt{-g_{00}}c^2 \Gamma^i_{jk} u^j u^k$.

Eq. (4.4):
$$F^i = ma^i = m\frac{d^2 x^i}{d\tau^2} = m\Gamma^i_{jk} \cdot \frac{dx^j}{d\tau} \cdot \frac{dx^k}{d\tau} $$
$$= m\sqrt{-g_{00}}c^2 \Gamma^i_{jk} u^j u^k$$

In the non-relativistic Newton approximation,

$$\alpha^i = \frac{d^2 x^i}{dt^2} = -\sqrt{-g_{00}}c^2 \Gamma^i_{jk} u^j u^k .$$

Eq. (4.5) yields a more simple equation from the condition of linear approximation, that is, weak-field, quasi-static, and slow motion (speed v << speed of light c: $u^0 \approx 1$):

$$\alpha^i = -\sqrt{-g_{00}} \cdot c^2 \Gamma^i_{00} .$$

4.2. Derivation of the Formula of Universal Gravitation

The formula of universal gravitational force was derived by Isaac Newton in 1665, and is phenomenologically used to explain observed facts and widely used in astronomical mechanics and spacecraft orbit calculations.

In general, the derivation of this universal gravitational equation is an indirect method because it uses Kepler's law.

The derivation of this universal gravitational equation should be directly derived as the gravitational force between two objects that are originally stationary, but unfortunately no such derivation method is found.

In this section, the law of universal gravitation is directly derived by a method different from the conventional method.

4.2.1. Conventional Derivation Method of the Formula of Universal Gravitation

Newton's law of universal gravitation is usually stated as that every objects attracts every other object in the universe with a force, that is directly proportional to the product of their masses and inversely proportional to the square of the distance between their centers.

As is well known, the equation for universal gravitation thus takes the form:

$$F = G\frac{m_1 m_2}{r^2},$$ (4.15)

where F is the gravitational force acting between two objects, m_1 and m_2 are the masses of the objects, r is the distance between the centers of their masses, and G is the gravitational constant.

By applying the equation of motion to Kepler's law, the law of universal gravitation is indirectly derived.

The following is a well-known conventional derivation method, but it is described for reference. The magnitude of the attractive force acting between the planet and the Sun is derived as follows (see Fig.4.1). To simplify the calculation, we assume that the planet is moving in a circular motion rather than an elliptical motion. The calculation for the elliptical orbit is more complicated than that for the circular orbit.

Then, according to Kepler's second law, this circular motion is a uniform circular motion: this attractive force is the centripetal force acting on a planet that moves circularly.

The centripetal force F can be used as:

$$F = mr\omega^2,$$ (4.16)

where F is centripetal force, m is the mass of the planet, r is the radius of the circular orbit (distance between the planet and the Sun), ω is the angular velocity, T is the rotation period.

Substituting $\omega = \dfrac{2\pi}{T}$ into Eq. (4.16), we get:

$$F = mr\left(\frac{2\pi}{T}\right)^2 = \frac{4\pi^2 mr}{T^2}.$$ (4.17)

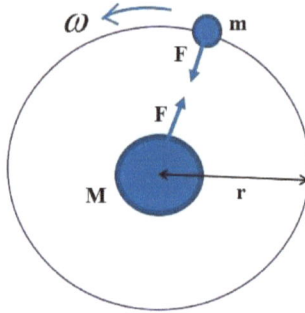

Fig. 4.1. Centripetal force F between objects M & m

Next, from Kepler's third law, we have

$$\frac{T^2}{r^3} = k.$$ (4.18)

Substituting Eq.(4.18) into Eq.(4.17), thus, we get:

$$F = \frac{4\pi^2 m}{kr^2} = G\frac{m}{r^2} \qquad (\frac{4\pi^2}{k} \Rightarrow G).$$ (4.19)

Eq. (4.19) means that the magnitude of the attractive force F acting on the planet is proportional to the mass of the planet and inversely proportional to the square of the distance from the Sun.

By the way, from the relation of action and reaction, the opposite direction and the same force F should act on the Sun. From the symmetry with F, this force is considered to be proportional to the mass M of the Sun and inversely proportional to the square of r.

Eq. (4.19) obtained above is an attractive force for the planet, but according to the law of action-reaction, the same magnitude of force F should be exerted on the Sun.

The magnitude of the attractive force acting on the Sun is also proportional to the mass of the Sun and inversely proportional to the square of the distance from the planet.

To satisfy above both states at the same time, the attractive force acting between the Sun and the planet is proportional to the mass of the Sun (M), proportional to the mass of the planet (m), and inversely proportional to the square of the distance between the sun and the planet.

Must be if this is expressed by an equation (the mass of the sun is M and the constant is G), the following equation is obtained:

$$F = G\frac{Mm}{r^2}. \tag{4.20}$$

In this way, the conventional derivation method uses centripetal force while the object is rotating, so derivation of the attractive force between two objects is indirect.

In addition, since the attraction between two objects is mentioned, the law of action-reaction will be brought up and must be applied.

What makes more important than anything else is that the attraction between the two objects must be directly derived between the two objects are originally stationary.

4.2.2. Another Derivation Method of the Formula of Universal Gravitation

As explained in Section 4.1, a massive body causes the curvature of space-time around it, and a free particle responds by moving along a geodesic line in that space-time. The path of free particle is a geodesic line in space-time and is given by the following geodesic equation;

$$\frac{d^2x^i}{d\tau^2} + \Gamma^i_{jk} \cdot \frac{dx^j}{d\tau}\frac{dx^k}{d\tau} = 0, \tag{4.21}$$

where $\Gamma^i{}_{jk}$ is Riemannian connection coefficient, τ is proper time, x^i is four-dimensional Riemann space, that is, three dimensional space ($x=x^1$, $y=x^2$, $z=x^3$) and one dimensional time ($w=ct=x^0$), c is the velocity of light. These four coordinate axes are denoted as x^i ($i=0, 1, 2, 3$).

As mentioned above, the acceleration (α) of curved space and its Riemannian connection coefficient (Γ^3_{00}) are given by:

$$\alpha = \sqrt{-g_{00}}\, c^2 \Gamma^3_{00} , \quad \Gamma^3_{00} = \frac{-g_{00,3}}{2g_{33}} , \tag{4.22}$$

where c: velocity of light, g_{00} and g_{33}: component of metric tensor, $g_{00,3}$: $\partial g_{00}/\partial x^3 = \partial g_{00}/\partial r$. We choose the spherical coordinates "$ct=x^0$, $r=x^3$, $\theta=x^1$, $\varphi=x^2$" in space-time. The acceleration α is represented by the equation both in the differential form and in the integral form. Practically, since the metric is usually given by the solution of gravitational field equation, the differential form has been found to be advantageous.

Now in general, the line element is described in:

$$\begin{aligned} ds^2 &= g_{ij} dx^i dx^j = g_{00}(dx^0)^2 + g_{33}(dx^3)^2 + g_{11}(dx^1)^2 + g_{22}(dx^2)^2 \\ &= g_{00}(cdt)^2 + g_{33}(dr)^2 + g_{11}r^2(d\theta)^2 + g_{22}r^2 \sin^2 \theta (d\varphi)^2 \end{aligned} \tag{4.23}$$

We choose the spherical coordinates "$ct=x^0$, $r=x^3$, $\theta=x^1$, $\varphi=x^2$" in space-time (see Fig. 4.2).

Next, let us consider External Schwarzschild Solution.

External Schwarzschild Solution is an exact solution of the gravitational field equation, which describes the gravitational field outside the spherically symmetric, static mass distribution.

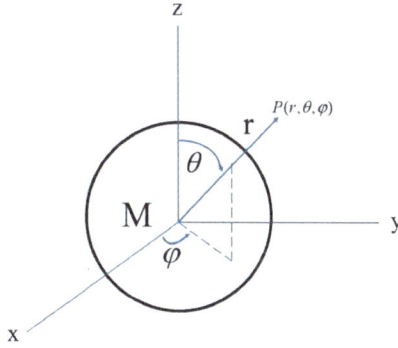

Fig. 4.2. Spherical coordinate system

The line element is obtained as follows:

$$ds^2 = -(1 - \frac{r_g}{r})c^2 dt^2 + \frac{1}{1 - \frac{r_g}{r}} dr^2 + r^2 (d\theta^2 + \sin^2 \theta d\varphi^2) . \qquad (4.24)$$

The metrics are given by:

$$g_{00} = -(1 - r_g/r), g_{11} = g_{22} = 1, g_{33} = 1/(1 - r_g/r),$$
$$and\ other\ g_{ij} = 0 . \qquad (4.25)$$

where r_g is the gravitational radius or Schwarzschild radius (i.e. $r_g = 2GM/c^2$).

Combining Eq. (4.25) with Eq. (4.22) yields:

$$\alpha = G \cdot \frac{M}{r^2}, (r_g \langle r) \quad , \qquad (4.26)$$

where G is gravitational constant and M is total mass.

Eq. (4.26) indicates the acceleration at a distance "r" from the center of the well-known the Earth mass M. The derivation process of this equation is shown below.

The force acting on a mass "m" located at a distance "r" from the center of the Earth mass M is:

$$F = ma = mG \frac{M}{r^2} = G \frac{Mm}{r^2} . \qquad (4.27)$$

Eq. (4.27) indicates a universal gravitational force acting on masses M and m that are stationary with respect to each other [5].

<Supplemental explanation for Eq. (4.26)>

Eq. (4.22): $\alpha = \sqrt{-g_{00}} \, c^2 \Gamma_{00}^3$, $\quad \Gamma_{00}^3 = \dfrac{-g_{00,3}}{2g_{33}}$.

The metrics are given by (Eq.4.25):

$$g_{00} = -\left(1 - \frac{r_g}{r}\right) = -1 + \frac{r_g}{r}, \, g_{33} = \frac{1}{1 - \dfrac{r_g}{r}}$$

Using Eq. (4.22),

$$g_{00,3} = \frac{\partial g_{00}}{\partial x^3} = \frac{\partial g_{00}}{\partial r} = \frac{\partial}{\partial r}\left(-1 + \frac{r_g}{r}\right) = -\frac{r_g}{r^2}$$

$$\Gamma_{00}^3 = \frac{-g_{00,3}}{2g_{33}} = \frac{r_g}{r^2} \cdot \frac{1 - \dfrac{r_g}{r}}{2} = \frac{r_g}{2r^2} - \frac{r_g^2}{2r^3} \approx \frac{r_g}{2r^2}$$

Since $r_g = \dfrac{2GM}{c^2}$ is the gravitational radius (Schwarzschild radius), then

$r_g \ll r$. Accordingly, the term of $\dfrac{r_g^2}{2r^3}$ is neglected.

Also, $g_{00} = -\left(1 - \dfrac{r_g}{r}\right) \approx -1$.

Acceleration (α) is obtained as Eq. (4.26):

$$\alpha = \sqrt{-g_{00}} \, c^2 \Gamma_{00}^3 = c^2 \Gamma_{00}^3 = c^2 \frac{r_g}{2r^2} = c^2 \frac{1}{2r^2} \frac{2GM}{c^2} = \frac{GM}{r^2}.$$

By the way, from $\Gamma_{nn}^m = -\dfrac{g_{nn,m}}{2g_{mm}}$, we used $\Gamma_{00}^3 = -\dfrac{g_{00,3}}{2g_{33}}$ (see Table.1).

$$\Gamma_{mmn} = (\Gamma_{nmm}) = -\frac{1}{2}g_{mm,n}$$

$$\Gamma_{mmm} = \frac{1}{2}g_{mm,m}$$

$$\Gamma_{mnn} = (\Gamma_{nmn}) = \frac{1}{2}g_{nn,m}$$

$$\Gamma_{mnm} = (\Gamma_{mmn}) = \frac{1}{2}g_{mm,n}$$

$$other \ \Gamma_{\mu\nu\lambda} = 0 \ (\because g_{\mu\nu} = 0 \ \mu \ne \nu)$$

$$\Gamma^{m}_{mn} = \Gamma^{m}_{nm} = \frac{g_{mm,n}}{2g_{mm}}$$

$$\Gamma^{m}_{mm} = \frac{g_{mm,m}}{2g_{mm}}$$

$$\Gamma^{m}_{nn} = -\frac{g_{nn,m}}{2g_{mm}}$$

$$other \ \Gamma^{\mu}_{\nu\lambda} = 0$$

$$R_{\mu\nu kl} = \frac{1}{2}(g_{ul,vk} - g_{vl,uk} - g_{uk,vl} + g_{vk,ul}) + \Gamma_{\beta ul}\Gamma^{\beta}_{vk} - \Gamma_{\beta uk}\Gamma^{\beta}_{vl}$$

4.3. Gravitational Acceleration on the Earth's Surface

4.3.1. Overview of the Linear Approximation of Weak Static Gravitational Fields

The acceleration α and major curvature R^{00} are given by

$$R^{00} = \frac{1}{2}g^{ij}h_{00,ij}, \quad \alpha = c^2\Gamma^{i}_{00} = \frac{1}{2}c^2 h_{00,i}, \tag{4.28}$$

respectively from the weak field approximation of the gravitational field equation.

Here, h_{00} is deviation between metric tensor g_{00} of curved space and Minkowski metric tensor η_{00} of flat space, that is,

$$g_{00} = \eta_{00} + h_{00} = -1 + h_{00}. \tag{4.29}$$

The notation of the symbol is as follows:

$$h_{00,ij} = \partial_i \partial_j h_{00} = \frac{\partial h_{00}}{\partial x^i \partial x^j}. \tag{4.30}$$

As is well known, the partial derivative $u_{i,j} = \partial_j u_i = \dfrac{\partial u_i}{\partial x^j}$ is not tensor

equation.

The covariant derivative $u_{i;j} = u_{i,j} - u_k \Gamma^k_{ij}$ is tensor equation and can be

carried over into all coordinate systems.

If the gravitational field is time-invariant, or static, and the gravitational

field is not very strong, Ricci tensor $R_{\mu\nu}$ is given by:

$$R_{\mu\nu} = \frac{1}{2}\left(\Box h_{\mu\nu} + h_{,\mu\nu} - h_{\mu\rho,\rho\nu} - h^\rho_{\nu,\mu\rho}\right)$$
$$= \frac{1}{2}\left(\Box h_{\mu\nu} + \partial_\mu \partial_\nu h - \partial_\rho \partial_\nu h_{\mu\rho} - \partial_\mu \partial_\rho h^\rho_\nu\right),$$

Where

$$\Box = \eta^{\mu\nu}\partial_\mu \partial_\nu = \nabla^2 - (\partial_0)^2. \tag{4.31}$$

Since all are static $h_{\mu\nu,0} = 0$ $(\partial_0 h_{\mu\nu} = 0)$ and now setting $\mu = \nu = 0$, this

component R_{00} is obtained

$$R_{00} = \frac{1}{2}\left(\nabla^2 h_{00} - (h_{00,0})^2 + h_{,00} - h_{0\rho,\rho 0} - h^\rho_{00,0\rho}\right) = \frac{1}{2}\nabla^2 h_{00}. \tag{4.32}$$

On the other hand,

$$g_{00} = \eta_{00} + h_{00} = -1 + h_{00} = -1 - \frac{2}{c^2}\phi. \tag{4.33}$$

As is well known, potential ϕ is

$$\phi = -\frac{GM}{R}, \tag{4.34}$$

where M is the Earth mass, R is the Earth radius, G is the Gravity constant.

We get,

$$h_{00} = -\frac{2}{c^2}\phi = -\frac{2}{c^2} \times -\frac{GM}{R} = \frac{2GM}{c^2 R}. \tag{4.35}$$

Then,

$$R_{00} = \frac{1}{2}\nabla^2 h_{00} = \frac{1}{2}\nabla^2\left(-\frac{2\phi}{c^2}\right) = -\frac{1}{c^2}\nabla^2\phi.$$ (4.36)

Curvature R_{00} can be described by the following approximation:

$$R_{00} = \frac{1}{2}\nabla^2 h_{00} = \frac{1}{2}\left(\frac{\partial^2}{\partial x^2}h_{00} + \frac{\partial^2}{\partial y^2}h_{00} + \frac{\partial^2}{\partial z^2}h_{00}\right),$$
$$= \frac{1}{2}\frac{d^2 h_{00}}{dx^2} = \frac{1}{2}h_{00}\frac{d^2}{dx^2} \approx \frac{1}{2}h_{00}/R^2$$ (4.37)

where x is toward the Earth center.

A similar result is obtained from Eq. (4.28) as:

$$R^{00} = 1/2 \cdot g^{ij}h_{00,ij} = 1/2 \cdot g^{33}\partial^2 h_{00}/\partial x^3\partial x^3 = 1/2 \cdot g^{33}\partial^2 h_{00}/\partial x\partial x$$
$$= 1/2 \cdot \partial^2 h_{00}/\partial r^2 \approx 1/2 \cdot h_{00}/R^2$$ (4.38)

The approximate expression for gravitational acceleration is:

$$\alpha = \frac{1}{2}c^2 h_{00,i} = \frac{1}{2}c^2 h_{00,x} = \frac{1}{2}c^2\frac{dh_{00}}{dx} \approx \frac{1}{2}c^2 h_{00}/R.$$ (4.39)

On the other hand, as described in previously (see Eq. (4.11)), the gravitational acceleration is also given by the following equation:

$$\alpha = \sqrt{-g_{00}}c^2\int_a^b R^{00}(r)dr.$$ (4.40)

Considering $g_{00} = -1$, substituting Eq. (4.37) or Eq. (4.38) into Eq. (4.40), we get:

$$\alpha = c^2\int_R^\infty \frac{1}{2}\frac{h_{00}}{r^2}dr = \frac{c^2}{2}\frac{h_{00}}{R}.$$ (4.41)

Eq. (4.41) coincides with Eq. (4.39), and the equation of gravitational acceleration expressed by Eq. (4.40) gives the physical meaning of mechanism of gravitation. This physical concept becomes clear in the next section.

Further, major curvature of Ricci tensor ($\mu = \nu = 0$) is calculated as follows:

$$R^{00} = g^{00}g^{00}R_{00} = -1\times-1\times R_{00} = R_{00}.$$ (4.42)

Here for convenience, raise the index and use it in the notation of R^{00} instead of R_{00}.

$$
R^{00} = \frac{1}{2} g^{ij} h_{00,ij} = \frac{1}{c^2} g^{ij} \left(\frac{1}{2} c^2 h_{00,i} \right)_{,j}
$$
$$
= \frac{1}{c^2} g^{ij} \alpha_{i,j} = \frac{1}{c^2} \alpha^j_{,j}
$$

(4.43)

where $h_{00,ij} = \dfrac{\partial h_{00}}{\partial x^i \partial x^j}$.

4.3.2. Gravitational Acceleration on the Earth

The accumulation of surface forces in a curved area of space from the Earth's surface R to the point at infinity (∞) gives the gravitational acceleration on the Earth's surface.

$$
\alpha = c^2 \int_R^\infty R^{00}(r) dr = c^2 \int_R^\infty \frac{1}{2} h_{00} \frac{1}{r^2} dr = -\frac{1}{2} c^2 h_{00} \left[\frac{1}{r} \right]_R^\infty
$$
$$
= -\frac{1}{2} c^2 h_{00} (0 - \frac{1}{R}) = \frac{1}{2} \frac{c^2 h_{00}}{R}
$$

(4.44)

From Eq. (4.35), the deviation h_{00} of the metric tensor from the flat space ($\eta_{00} = -1$) on the Earth's surface becomes:

$$
h_{00} = \frac{2GM}{c^2 R}.
$$

(4.45)

The curvature of the space is (see Eq. 4.37):

$$
R_{00} = \left(\frac{1}{2} h_{00} \right) / R^2.
$$

(4.46)

The gravitational acceleration is (see Eq. 4.39):

$$
\alpha = (\frac{1}{2} c^2 h_{00}) / R.
$$

(4.47)

Substitute the values of the Earth radius R=6.378×10³km, GM=3.986×10⁵km³/s², c=3×10⁵km, we get the following values respectively:

$$h_{00} = \frac{2GM}{c^2 R} = \frac{2 \times 3.986 \times 10^5}{(3 \times 10^5)^2 \times 6.378 \times 10^3} = \frac{2 \times 3.986 \times 10^5}{9 \times 6.378 \times 10^{13}} = 1.389 \times 10^{-9}$$

$$R_{00} = \left(\frac{1}{2} h_{00}\right) \times \frac{1}{R^2} = \frac{1}{2} \times 1.389 \times 10^{-9} \times \frac{1}{(6.378 \times 10^3)^2} = 1.71 \times 10^{-2} \times 10^{-9} \times 10^{-6}$$

$$= 1.71 \times 10^{-17} (1/km)^2 = 1.71 \times 10^{-23} (1/m^2)$$

$$\alpha = \left(\frac{1}{2} c^2 h_{00}\right) \times \frac{1}{R} = \frac{1}{2} \times (3 \times 10^5)^2 \times 1.389 \times 10^{-9} \times \frac{1}{6.378 \times 10^3}$$

$$= \frac{1}{2} \times 9 \times 1.389 \times \frac{1}{6.378} \times 10^{10} \times 10^{-9} \times 10^{-3} = 0.98 \times 10^{-2} km/s^2 = 9.8m/s^2$$

In this way, the following approximate values can be obtained:
The amount of displacement of the space on the Earth's surface:

$$h_{00} = 1.389 \times 10^{-9}.$$

Curvature of space on the Earth's surface: $R_{00} = 1.71 \times 10^{-23}/m^2$.

Gravitational acceleration on the Earth's surface $\alpha = 9.8m/s^2$.

Next, a description will be given using the drawings.

Fig. 4.3 shows a concentric curved space area around the Earth. Distance of radius R from the center of the Earth is the surface of the Earth.

At a distance from the Earth to an infinite point, the space becomes flat space without being affected by the gravitation of the Earth. The point at infinity is indicated by a symbol ∞ and a dotted line.

The accumulation of surface forces in a curved area of space from the Earth's surface R to the point at infinity (∞) gives the gravitational acceleration on the Earth's surface, i.e., $\alpha_R = 9.8m/s^2$.

$$u = c^2 \int_R^\infty R^{00}(r)dr = c^2 \int_R^{+\infty} \frac{1}{2} h_{00} \frac{1}{r^2} dr = -\frac{1}{2} c^2 h_{00} \left[\frac{1}{r}\right]_R^{+\infty}$$

$$= -\frac{1}{2} c^2 h_{00} (0 - \frac{1}{R}) = \frac{1}{2} \frac{c^2 h_{00}}{R}$$

(4.48)

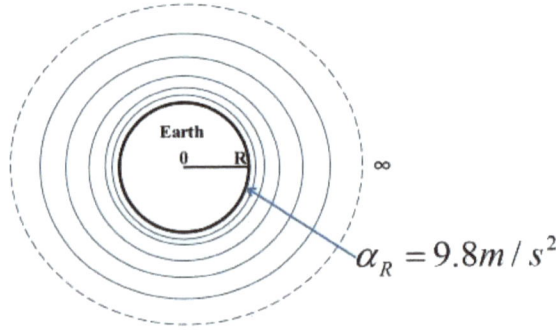

$$\alpha = c^2 \int_R^\infty R^{00}(r)\,dr = c^2 \int_R^\infty \frac{1}{2}h_{00}\frac{1}{r^2}\,dr = -\frac{1}{2}c^2 h_{00}\left[\frac{1}{r}\right]_R^\infty = -\frac{1}{2}c^2 h_{00}(0-\frac{1}{R}) = \frac{1}{2}\frac{c^2 h_{00}}{R}$$

Fig. 4.3. Mechanism of gravitational acceleration generation on the Earth's surface

As well, Fig.4.4 shows a concentric curved space area around the Earth. Distance of radius R from the center of the Earth is the surface of the Earth. Here consider the gravitational acceleration at a height **h** away from the Earth's surface.

At a distance from the Earth to an infinite point, the space becomes flat space without being affected by the gravitation of the Earth. The point at infinity is indicated by a symbol ∞ and a dotted line.

The accumulation of surface forces in a curved area of space from the Earth's surface R+**h** to the point at infinity (∞) gives the gravitational acceleration at the Earth's height **h**, i.e., $\alpha_{R+h} < \alpha_R = 9.8m/s^2$.

$$\alpha = c^2 \int_{R+h}^\infty R^{00}(r)\,dr = c^2 \int_{R+h}^\infty \frac{1}{2}h_{00}\frac{1}{r^2}\,dr = -\frac{1}{2}c^2 h_{00}\left[\frac{1}{r}\right]_{R+h}^\infty$$
$$= -\frac{1}{2}c^2 h_{00}(0-\frac{1}{R+h}) = \frac{1}{2}\frac{c^2 h_{00}}{(R+h)}$$

(4.49)

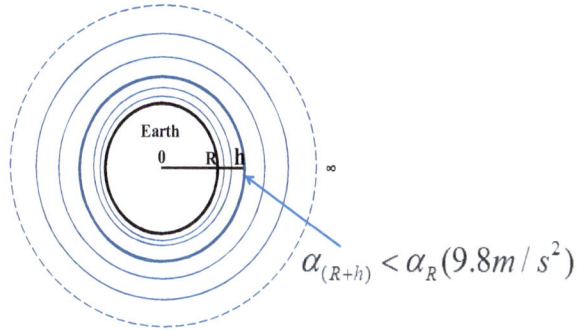

$$\alpha = c^2 \int_{R+h}^{\infty} R^{00}(r) \, dr = c^2 \int_{R+h}^{\infty} \frac{1}{2} h_{00} \frac{1}{r^2} \, dr = -\frac{1}{2} c^2 h_{00} \left[\frac{1}{r} \right]_{R+h}^{\infty} = -\frac{1}{2} c^2 h_{00} (0 - \frac{1}{R+h}) = \frac{1}{2} \frac{c^2 h_{00}}{(R+h)}$$

Fig. 4.4. Mechanism of gravitational acceleration generation on the Earth's surface height h

As described above, although the spatial curvature at the surface of the Earth is very small value, i.e., $1.71 \times 10^{-23} (1/m^2)$, it is enough value to produce 1G (9.8 m/s²) acceleration.

5. APPLICATION TO OUTER SPACE

Comparing the space on the ground and the space in outer space, although there seems to be no difference, obviously a different phenomenon occurs. Simply put, an object moves radially inward, that is, drops straight down on the Earth, but in the outer space, the object floats and does not move.

The difference between the two phenomena can be explained by whether space is curved or not. In essence, the existence of spatial curvature and curved extent region determine whether the object drops straight down or not.

Although the spatial curvature at the surface of the Earth is very small value, i.e., $1.71 \times 10^{-23} (1/m^2)$, it is enough value to produce 1G (9.8 m/s^2) acceleration.

Conversely, the spatial curvature in the outer space is zero, therefore any acceleration is not produced. Accordingly, if the spatial curvature of a localized area containing object is controlled to the curvature of $1.71 \times 10^{-23} (1/m^2)$ with a sufficiently large curved space area, the object moves and receives 1G acceleration in the outer space. Of course, we are required to control both the magnitude of the curvature and the size of the curved space area.

So how can we curve the space artificially? As a matter of fact, space curvature is generated not only by mass energy but also by electromagnetic energy.

From General Relativity, the major component of curvature of space R^{00} can be produced by not only mass density but also the magnetic field B as follows:

$$R^{00} = \frac{4\pi G}{\mu_0 c^4} \cdot B^2 = 8.2 \times 10^{-38} \cdot B^2 \quad (B \text{ in Tesla}), \tag{5.1}$$

where $\mu_0 = 4\pi \times 10^{-7} (H/m)$, $\varepsilon_0 = 1/(36\pi) \times 10^{-9} (F/m)$, $G = 6.672 \times 10^{-11} (N \cdot m^2/kg^2)$,

$c = 3 \times 10^8 (m/s)$,

B is a magnetic field in Tesla and R^{00} is a major component of spatial curvature $(1/m^2)$.

Eq.(5.1) indicates that the major component of spatial curvature can be controlled by magnetic field B. In case that the intensities of the magnetic field B and the electric field E are equal, the value of $(1/2 \cdot \varepsilon_0 E^2)$ is about seventeen figures smaller than the value of $(B^2/2\mu_0)$. As a consequence, the electric field only negligibly contributes to the spatial curvature as compared with the magnetic field.

The relationship between curvature and magnetic field was derived by Minami and introduced it in 16th International Symposium on Space Technology and Science (1988) [8]. Minami proposed a hypothesis for mechanical property of space-time in 1988. A primary motive was to research in the realm of space propulsion theory using the substantial physical structure of space-time based on this hypothesis.

The concepts and necessary mathematical formulas in Chapters 3 and 4 described in this book were derived in this paper in 1988, and this time, the description focuses on the viewpoint of the gravity generation mechanism.

Please refer to **APPENDIX A** for Eq. (5.1) derivation [8, 9].

6. GENERATION OF TORSION FIELD BY ROTATION

In this chapter, it is not directly related to the generation of gravity, but it will be introduced as an interesting theme in the future.

Torsion field or twisted field generation by rotation and its asymmetric gravitational field are described.

General Relativity does not target twisted fields or torsion fields. This is equivalent to taking a geodetic coordinate system, which are locally flat coordinates at any point in space-time. While, as is well known, torsion tensors do not disappear in some gravity theories (Einstein-Cartan theory, etc.). Riemann space given the Levi-Civita parallelism is a special Euclidean connection space (i.e., non-torsion). That is, the connection coefficients are taken to be symmetric $\Gamma^i_{jk} = \Gamma^i_{kj}$. In General Relativity, since the torsion tensor is taken to disappear, the connection coefficients are taken to be symmetric $\Gamma^i_{jk} = \Gamma^i_{kj}$. Therefore, the acceleration has the same value regardless of the direction of rotation. This means that the connections are symmetric, as General Relativity takes the torsion tensor $S^i_{jk} = \Gamma^i_{jk} - \Gamma^i_{kj}$ to disappear, i.e.,

$$S^i_{jk} = 0.$$

6.1. Concept of Torsion Field

As a geometric interpretation of the torsion field, an infinitesimal parallelogram composed of infinitesimal translations of vectors gives a valid concept.

As a concept, the torsion indicates a state in which a parallelogram formed by infinitesimal translation of a vector on two sides does not close. The parallelogram closes only when the torsion is zero.

Torsion tensor $S^i_{jk} = \Gamma^i_{jk} - \Gamma^i_{kj}$ measures the asymmetric part of the connection coefficients $\Gamma^i_{jk}, \Gamma^i_{kj}$.

The tangent vector $\dfrac{du^i}{ds}$ of a geodesic line $u^i(s)$ moves parallel along the geodesic line. Conversely, if the tangent vector $\dfrac{du^i}{ds}$ of a curve line $u^i(s)$ moves parallel along that curve line, then the curve line $u^i(s)$ is a geodesic line.

For an equation that represents a vector $\lambda^i(s)$ moving parallel along a curve line $u^i(s)$, that is,

$$\frac{d\lambda^i}{ds} + \Gamma^i_{jk}\lambda^j \frac{du^k}{ds} = 0, \tag{6.1}$$

if λ^i is one vector, $\delta\lambda^i = d\lambda^i + \Gamma^i_{jk}\lambda^j du^k$ also becomes one vector and is called the covariant derivative of the vector.

The vector λ^i at the point u^i and the vector $\lambda^i + d\lambda^i$ at the point $u^i + du^i$ are defined to be parallel when $\delta\lambda^i = 0$.

Therefore, when the vector $\lambda^i(s)$ defined at each point on the curve line $u^i(s)$ is satisfying a differential equation $\dfrac{\delta\lambda^i}{ds} = \dfrac{d\lambda^i}{ds} + \Gamma^i_{jk}\lambda^j \dfrac{du^k}{ds} = 0$, then, it is parallel along the curve line [10].

In other words, from $\delta\lambda^i = d\lambda^i + \Gamma^i_{jk}\lambda^j du^k = 0$, we get:

$$d\lambda^i = -\Gamma^i_{jk}\lambda^j du^k. \tag{6.2}$$

Let's consider the following figure (Fig.6.1) using these results.

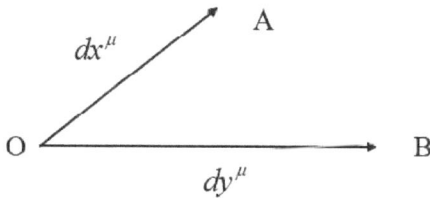

Fig. 6.1. Triangle AOB

For a triangle AOB with infinitesimal sides dx^μ and dy^μ, consider an infinitesimal translation of a vector dx^μ from O to B, and consider an infinitesimal translation of a vector dy^μ from O to A [10, 11].

At this time, two infinitesimal vectors $dx^\mu - \Gamma^\mu_{\lambda\sigma} dx^\lambda dy^\sigma$ and $dy^\mu - \Gamma^\mu_{\lambda\sigma} dy^\lambda dx^\sigma$ are obtained, and a new parallelogram is formed by these four vectors. But the parallelogram does not close as shown in Fig. 6.2.

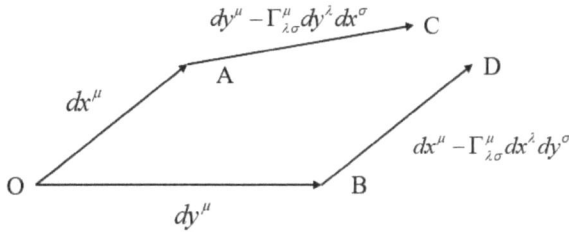

Fig. 6.2. Parallelogram

In order for this figure to close and become a parallelogram, the vector CD = 0.

$$OC\text{-}OD=0 \quad \Rightarrow \quad OA+AC-(OB+BD)=0,$$

that is,

$$dx^\mu + dy^\mu - \Gamma^\mu_{\lambda\sigma} dy^\lambda dx^\sigma - (dy^\mu + dx^\mu - \Gamma^\mu_{\lambda\sigma} dx^\lambda dy^\sigma) =$$

$$dx^\mu + dy^\mu - \Gamma^\mu_{\sigma\lambda} dy^\sigma dx^\lambda - (dy^\mu + dx^\mu - \Gamma^\mu_{\lambda\sigma} dx^\lambda dy^\sigma).$$

(Here, for convenience, the index λ,σ of the connection coefficient of the third term is interchanged)

$$= dx^\mu + dy^\mu - \Gamma^\mu_{\sigma\lambda} dx^\lambda dy^\sigma - (dy^\mu + dx^\mu - \Gamma^\mu_{\lambda\sigma} dx^\lambda dy^\sigma)$$

$$= \Gamma^\mu_{\lambda\sigma} dx^\lambda dy^\sigma - \Gamma^\mu_{\sigma\lambda} dx^\lambda dy^\sigma = \left(\Gamma^\mu_{\lambda\sigma} - \Gamma^\mu_{\sigma\lambda}\right) dx^\lambda dy^\sigma. \qquad (6.3)$$

To do this, if the connection coefficient $\Gamma^{\mu}_{\lambda\sigma} = \Gamma^{\mu}_{\sigma\lambda}$ is symmetric, the parallelogram is closed and the torsion tensor $S^{\mu}_{\lambda\sigma} = \Gamma^{\mu}_{\lambda\sigma} - \Gamma^{\mu}_{\sigma\lambda} = 0$.

The fact that the field of space is twisted means that the parallelogram does not close even if the vector is translated. Zero torsion corresponds to the closing of an infinitesimal parallelogram with translation [10, 11].

Namely, if torsion tensor $S^{\mu}_{\lambda\sigma} = \left(\Gamma^{\mu}_{\lambda\sigma} - \Gamma^{\mu}_{\sigma\lambda} \right) dx^{\lambda} dy^{\sigma} \neq 0$, then the connection coefficient becomes $\Gamma^{\mu}_{\lambda\sigma} \neq \Gamma^{\mu}_{\sigma\lambda}$.

Therefore, acceleration generated in a curved space has different magnitudes because the connection coefficients are different in a twisted field:

$$\alpha^{\mu} = \frac{d^2 x^{\mu}}{d\tau^2} = -\Gamma^{\mu}_{\lambda\sigma} \cdot \frac{dx^{\lambda}}{d\tau} \cdot \frac{dx^{\sigma}}{d\tau} = -\sqrt{-g_{00}} c^2 \Gamma^{\mu}_{\lambda\sigma} u^{\lambda} u^{\sigma}, \tag{6.4}$$

where u^{λ}, u^{σ} are the four velocity, $\Gamma^{\mu}_{\lambda\sigma}$ is the Riemannian connection coefficient, and τ is the proper time.

The following figure (Fig.6.3) can be considered as one image of the concept of a torsion field. It is a picture like a viscous fluid of a vortex.

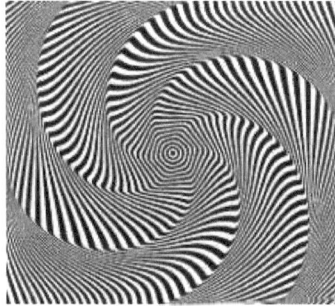

Fig. 6.3. Image of a torsion field

6.2. Effect of Rotation on Space-Time

In this section, we consider the effect of rotation on the surrounding space-time.

Regarding the influence of the rotation of the object on the surrounding space-time, the Kerr metric will be the target, here the useful method described in Section 4.2.2 will be described again.

6.2.1. Derivation of the Acceleration of Static Mass

Now in general, the line element is described in:

$$ds^2 = g_{ij}dx^i dx^j = g_{00}(dx^0)^2 + g_{33}(dx^3)^2 + g_{11}(dx^1)^2 + g_{22}(dx^2)^2$$
$$= g_{00}(cdt)^2 + g_{33}(dr)^2 + g_{11}r^2(d\theta)^2 + g_{22}r^2 \sin^2 \theta(d\varphi)^2 \qquad (6.5)$$

We choose the spherical coordinate system "ct=x⁰, r=x³, θ=x¹, φ=x²" in space-time (see Fig.6.4).

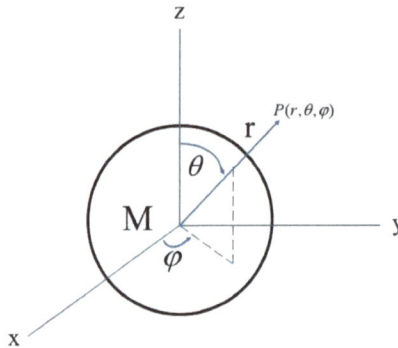

Fig. 6.4. Spherical coordinate system

As already mentioned in Section 4.2.2, we will introduce it again here as an example of use. Let us consider External Schwarzschild Solution.

External Schwarzschild Solution is an exact solution of the gravitational field equation, which describes the gravitational field outside the spherically symmetric, static mass distribution.

The line element is obtained as follows:

$$ds^2 = -(1-\frac{r_g}{r})c^2 dt^2 + \frac{1}{1-\frac{r_g}{r}}dr^2 + r^2(d\theta^2 + \sin^2 \theta d\varphi^2). \qquad (6.6)$$

The metrics are given by:

$$g_{00} = -(1 - r_g/r), g_{11} = g_{22} = 1, g_{33} = 1/(1 - r_g/r),$$
$$and \ other \ g_{ij} = 0 .$$

(6.7)

where r_g is the gravitational radius (i.e., $r_g = 2GM/c^2$).

Combining Eq. (6.7) with Eq. (4.22) yields:

$$\alpha = G \cdot \frac{M}{r^2}, (r_g \langle r) \quad ,$$

(6.8)

where G is gravitational constant and M is total mass.

Eq. (6.8) indicates the acceleration at a distance "r" from the center of the Earth mass M.

The force acting on a mass "m" located at a distance "r" from the center of the Earth mass M is:

$$F = m\alpha = mG\frac{M}{r^2} = G\frac{Mm}{r^2} .$$

(6.9)

Eq. (6.9) indicates a universal gravitational force acting on masses M and m that are stationary with respect to each other [5].

6.2.2. Derivation of Kerr Rotating Mass Solution

Kerr metric targets an axisymmetric star that rotates constantly like black holes.

The line element in Kerr space-time is obtained as follows:

$$ds^2 = -\left(1 - \frac{r_g r}{r^2 + h^2 \cos^2 \theta}\right) c^2 dt^2 - \frac{2r_g h \sin^2 \theta}{r^2 + h^2 \cos^2 \theta} rcdtd\varphi + \frac{r^2 + h^2 \cos^2 \theta}{r^2 - r_g r + h^2} dr^2$$
$$+ \left(1 + \frac{h^2}{r^2}\cos^2 \theta\right) r^2 d\theta^2 + \left(1 + \frac{h^2}{r^2} + \frac{r_g h^2 r \sin^2 \theta}{r^4 + r^2 h^2 \cos^2 \theta}\right) r^2 \sin^2 \theta d\varphi^2$$

(6.10)

M is the mass of object and J is the angular momentum. The metrics outside of spinning mass are given by:

$$g_{00} = -\left(1 - \frac{r_g r}{r^2 + h^2 \cos^2 \theta}\right), \quad g_{33} = \frac{r^2 + h^2 \cos^2 \theta}{r^2 - r_g r + h^2} ,$$

$$g_{11} = 1 + \frac{h^2}{r^2}\cos^2 \theta , \quad g_{22} = 1 + \frac{h^2}{r^2} + \frac{r_g h^2 r \sin^2 \theta}{r^4 + r^2 h^2 \cos^2 \theta} ,$$

(6.11)

where $h = J/Mc$ ($J = angular \ momentum$), $r_g = 2GM/c^2$.

Eq. (6.11) reduces to the Schwarzschild solution if the angular momentum "J" is zero.

On the while, the acceleration (α) of curved space and its Riemannian connection coefficient (Γ_{00}^3) are given by:

$$\alpha = \sqrt{-g_{00}}\, c^2 \Gamma_{00}^3 \ , \qquad \Gamma_{00}^3 = \frac{-g_{00,3}}{2g_{33}} \ , \tag{6.12}$$

where c=speed of light, g_{00} and g_{33}=component of metric tensor, and

$$g_{00,3} = \partial g_{00}/\partial x^3 = \partial g_{00}/\partial r \ .$$

We choose the spherical coordinates "ct=x^0, r=x^3, θ=x^1, φ=x^2" in space-time.

Combining Eq. (6.11) with Eq. (6.12) yields:

$$\alpha = G \cdot \frac{M}{r^2} \cdot \frac{(1-h^2\cos^2\theta/r^2)}{(1+h^2\cos^2\theta/r^2)^3} < G \cdot \frac{M}{r^2} \ , \ (r_g < r, h^2 < r^2). \tag{6.13}$$

*< For the derivation process of Eq. (6.13), see **APPENDIX B** >*

Eq. (6.13) indicates that **the rotation weakens the gravitational acceleration.**

6.3. Torsion Field Generation due to Rotation

The only effect of a static black hole on space-time is the distortion of space (curved space) due to mass, but when a black hole rotates, the surrounding space-time becomes distorted due to not only mass but also accompanying rotation. The surrounding space-time also rotates with the black hole. The space-time around the black hole is a large spiral that flows into the black hole at the same time. Such a space-time rotation effect is called a frame dragging [12, 13].

6.3.1. Frame-dragging by Rotating Mass Body

When a mass body rotates at high speed in space, a phenomenon called frame-dragging (Lense-Thirring effect) is well known among the space-time phenomena predicted by General Relativity. It is about a rotating celestial body, which has the effect of spirally pulling the surrounding local inertial frame into it. The space-time structure around the rotating source of gravity is represented by the Kerr metric. It is a spiral rotation that increases rapidly as it approaches the gravity source, and it can be interpreted that the inertial system is dragged by the rotation of the gravity source like a spiral viscous fluid. Due to this drag, the object does not fall straight to the source of gravity, but spins and falls. This is a phenomenon peculiar to General Relativity called "frame dragging".

Often targeted at the frame-dragging effect of rotating black holes, this phenomenon applies not only to black holes, but also to masses such as the Earth and even smaller gyroscopes. The space-time around a rotating object is dragged by the rotation of the object and rotates in the same direction. And, the space-time around the rotating object is inversely proportional to the cube of the distance (r) from the object, and rotates around the object at an angular velocity (ω) proportional to the angular momentum (J) of the object. Here, x is the x component of force.

$$\omega = \frac{G}{r^3}\left[J - \frac{3}{r^2}(Jx)x\right].$$ (6.14)

The dragging effect of the frame-dragging on the Earth's rotation has already been confirmed on the Gravity Probe B satellite launched by NASA and Stanford University in 2004.

Gravity Probe B is activated in polar orbit and is equipped with four gyroscopes, and the drag effect of the inertial system is verified by measuring the deviation of the rotation axis of these gyroscopes. As a result, the dragging effect of the space-time due to the rotation of the Earth has been confirmed with an accuracy of 20% from the observation of the precession of the gyroscope. This phenomenon is the dragging effect of space-time due to the rotation of the Earth, and is not applied only to the ergosphere such as black holes [12].

As already explained in Section 6.2.2, when the mass rotates at high speed in space, the gravitational acceleration decreases due to the Kerr solution. Gravitational acceleration decreases regardless of the direction of rotation, that is, clockwise or counterclockwise rotation.

However, this is a space with zero torsion, and there is no torsion field in the space. Therefore, the connection coefficients are symmetric. In other words, in General Relativity, the torsion tensor is taken to disappear, which means that the connection is symmetric. This is equivalent to taking a geodetic coordinate system (where $\Gamma^{\mu}_{\lambda\sigma}$ are all "0" locally), which are locally flat coordinates at any point in space-time.

However, if it is assumed that the space field is twisted from the beginning, there is a difference in the effect that the local inertial system around the clockwise or counterclockwise rotations is drawn into the spiral. In this case, since the torsion tensor is not zero, the connection coefficient becomes asymmetric and the value of the acceleration field becomes different. Torsion tensor measures the asymmetric part of the connection factor, that is,

$$S^{\mu}_{\lambda\sigma} = \Gamma^{\mu}_{\lambda\sigma} - \Gamma^{\mu}_{\sigma\lambda}.$$

6.3.2. Torsion Field and Direction of Rotation

As has been already mentioned, the rotation of an object has some effect on the surrounding space as the following.

(1) *Rotation weakens the gravitational acceleration* (see 6.2.2 Derivation of Kerr Rotating Mass Solution):

Same reduction in gravitational acceleration occurs regardless of left or right rotation.

In General Relativity, the connection coefficients are taken to be symmetric $\Gamma^{i}_{jk} = \Gamma^{i}_{kj}$. Therefore, the acceleration has the same value regardless of the direction of rotation. This means that the connections are symmetric, as General Relativity takes the torsion tensor $S^{i}_{jk} = \Gamma^{i}_{jk} - \Gamma^{i}_{kj}$ to disappear, i.e.,

$$S^{i}_{jk} = 0.$$

It is rotating, but there is no torsion field.

(2) Frame-dragging (Lense-Thirring effect) *(see 6.3.1 Frame-dragging by Rotating Mass Body)*

This phenomenon occurs in the same way regardless of the direction of left rotation and right rotation.

It is rotating, but there is no torsion field.

The above two items (1) and (2) cannot explain the following question.

"Does the gravitational acceleration differ depending on the direction of rotation of the object (clockwise and counterclockwise)?"

If so, the rotation will create a torsion field.

Because, as is already explained, if torsion tensor

$$S^\mu_{\lambda\sigma} = \left(\Gamma^\mu_{\lambda\sigma} - \Gamma^\mu_{\sigma\lambda}\right) dx^\lambda dy^\sigma \neq 0,$$

then the connection coefficient becomes $\Gamma^\mu_{\lambda\sigma} \neq \Gamma^\mu_{\sigma\lambda}$. Therefore, acceleration generated in a curved space has different magnitudes because the connection coefficients are different in a torsion field: $\Gamma^\mu_{\lambda\sigma} \neq \Gamma^\mu_{\sigma\lambda}$.

From

$$\alpha^\mu = \frac{d^2 x^\mu}{d\tau^2} = -\Gamma^\mu_{\lambda\sigma} \cdot \frac{dx^\lambda}{d\tau} \cdot \frac{dx^\sigma}{d\tau} = -\sqrt{-g_{00}} c^2 \Gamma^\mu_{\lambda\sigma} u^\lambda u^\sigma, \qquad (6.15)$$

we get:

$$\alpha^\mu(R) = -\sqrt{-g_{00}} c^2 \Gamma^\mu_{\lambda\sigma} u^\lambda u^\sigma \neq -\sqrt{-g_{00}} c^2 \Gamma^\mu_{\sigma\lambda} u^\lambda u^\sigma = \alpha^\mu(L), \qquad (6.16)$$

where u^λ, u^σ are the four velocity, $\Gamma^\mu_{\lambda\sigma}$ is the Riemannian connection coefficient, and τ is the proper time.

Namely, there occur the accelerations of different strength ($\alpha^\mu(R) \neq \alpha^\mu(L)$) in both rotations of opposite directions due to different connection coefficients ($\Gamma^\mu_{\lambda\sigma} \neq \Gamma^\mu_{\sigma\lambda}$).

Does the rotation create a torsion field? Or, does the twisting direction of the torsion field determine the difference in phenomenon depending on the direction of rotation?

The deciding factor may be whether to target the global space due to rotation or the local space due to rotation. Furthermore, it depends on whether the rotation is high speed rotation.

After all, if there is a possibility, it can be considered that a torsion field is generated in the space of the local region around the fast-rotating object.

The above is expected to be examined in the future.

CONCLUSION

Assuming that space is an infinite continuum, a mechanical concept of space became identified. Space can be considered as a kind of transparent elastic field. The pressure field derived from the geometrical structure of space is newly obtained by applying both continuum mechanics and General Relativity to space. As a result, a fundamental concept of space-time is described that focuses on theoretically innate properties of space including strain and curvature.

The mass on the Earth will not be pulled by the Earth and fall, but will be pushed and fall in the direction of the Earth due to the pressure of the field in the curved space area around the Earth. Although the spatial curvature at the surface of the Earth is very small value, i.e., $1.71 \times 10^{-23} (1/m^2)$, it is enough value to produce 1G (9.8 m/s^2) acceleration.

This concept can be also applied to the gravity effect of stars, stars in the galactic universe, planets, etc.

As the essence of gravitation, the cause and mechanism of gravitation can be explained from a single concept of the pressure field in space. Gravitation (Gravity) can be explained as a pressure field induced by the curvature of space.

As is well known, there are four kinds of forces in the universe: strong force, electromagnetic force, weak force, and gravity. Since the space-time is created at the beginning after the birth of the universe, gravity will be generated compared to other forces first.

Except for gravity, quantum sources have been discovered, and those forces are described in quantum theory. The underlying quantum graviton for gravity is not found, and gravity has not been described successfully in quantum theory.

Some basic concepts of quantum theory and General Relativity are incompatible. In quantum theory, one time must be set in all flat space-time to define a state. Meanwhile, in the General Relativity, time is not defined globally over the entire space-time and can be introduced only locally, which makes it difficult to connect with quantum theory. It is not possible to renormalize even if it is locally limited, and it is still unsolved, including infinite divergence.

In conclusion, **Gravitation was a pushing force, not a pulling force.**

The apples on the Earth will not be pulled by the Earth and fall, but will be pushed and fall in the direction of the Earth due to the pressure of the field in the curved space area around the Earth.

The book "GRAVITATION - Its Cause and Mechanism –" disclosed a clear principle of operation for an object falling on the Earth. Furthermore, it also brought about different derivation method of the law of universal gravitation from the conventional method.

Especially the following equation $\alpha = \sqrt{-g_{00}}\, c^2 \int_a^b R^{00}(r)dr$ gives the physical meaning of mechanism of gravitation.

ACKNOWLEDGEMENTS

We sincerely thanks for their excellent books continuously guiding this research [14, 15, 16, 17].

REFERENCES

[1] Williams, C. (Editor); Minami, Y. (Chap.3); et al. *Advances in General Relativity Research*, Nova Science Publishers, 2015.

[2] Minami, Y., "Continuum Mechanics of Space Seen from the Aspect of General Relativity — An Interpretation of the Gravity Mechanism", *Journal of Earth Science and Engineering* 5, 2015: 188-202.

[3] Minami, Y., "Gravitational Effects Generated by the Curvature of Space on the Earth's Surface", *Journal of Scientific and Engineering Research*, 2020, 7(3):1-15.

[4] Minami, Y., *Mechanism of GRAVITY Generation—why apples fall—*, published in May. 1, 2020, (LAMBERT Academic Publishing); https://www.morebooks.shop/gb/search?page=4&per_page=16&q=Yoshinari+Minami&search_term=Yoshinari+Minami&utf8=%E2%9C%93

[5] Minami, Y.,"Another Derivation Method Of The Formula Of Universal Gravitation", Science and Technology Publishing (SCI & TECH), Vol.4 Issue 6: 291-296, 2020.

[6] Minami, Y., "Gravity and Acceleration Produced in a Curved Space", Science and Technology Publishing (SCI & TECH), Vol.4 Issue 8: 450-460, 2020.

[7] Minami, Y., "Issues Left in General Relativity", Science and Technology Publishing (SCI & TECH), Vol.4 Issue 9: 487-491, 2020.

[8] Minami, Y., "Space Strain Propulsion System", *Proceedings of 16th International Symposium on Space Technology and Science*, 1988, 125-36.

[9] Minami, Y., "On the Possibility of Gravity Control by Magnetic Field", Scholars Journal of Engineering and Technology (SJET), 2018.

[10] Yano, K., *Geometry of connection*, Morikita Shuppan, 1968.

[11] Nash, C., Sen, S., *Topology and Geometry for Physicists*, McGraw-Hill, 1989.

[12] Futamase, T., *Theory of Relativity*, Asakura Publishing Co., Ltd., 2020.

[13] Minami, Y., "Gravitational Acceleration due to the Torsion Field of Space-Time: Weight Reduction on A Gyroscope's Right Rotation", American Journal of Sciences and Engineering Research, Vol.4, Issue 1, 2021, 15-27.

[14] Flügge, W., *Tensor Analysis and Continuum Mechanics*, New York: Springer-Verlag, 1972.

[15] Fujii, Y., *Space-Time and Gravity*, Sangyo Tosho Publishing Co., Ltd., Tokyo, 1979.

[16] Uchiyama, R., *Theory of Relativity*, Iwanami Shoten, Publishers, Tokyo, 1984.

[17] Hirakawa, M., *Relativity*, Kyoritsu Shuppan Co., Ltd., 1986.

[18] Wolfgang, Pauli., *Theory of Relativity*, Dover Publications, Inc., New York, 1981.

APPENDICES

APPENDIX A: Curvature Control by Magnetic Field

Let us consider the electromagnetic energy tensor M^{ij}. In this case, the solution of metric tensor g_{ij} is found by

$$R^{ij} - \frac{1}{2} \cdot g^{ij} R = -\frac{8\pi G}{c^4} \cdot M^{ij} . \tag{A.1}$$

Eq. (A.1) determines the structure of space due to the electromagnetic energy.

Here, if we multiply both sides of Eq. (A.1) by g_{ij}, we obtain:

$$g_{ij}\left(R^{ij} - \frac{1}{2} \cdot g^{ij} R \right) = g_{ij} R^{ij} - \frac{1}{2} \cdot g_{ij} g^{ij} R = R - \frac{1}{2} \cdot 4R = -R , \tag{A.2}$$

$$g_{ij}\left(\frac{-8\pi G}{c^4} \cdot M^{ij} \right) = -\frac{8\pi G}{c^4} \cdot g_{ij} M^{ij} = \frac{-8\pi G}{c^4} \cdot M^i_i = \frac{-8\pi G}{c^4} M . \tag{A.3}$$

The following equation is derived from Eqs.(A.2) and (A.3)

$$R = \frac{8\pi G}{c^4} \cdot M . \tag{A.4}$$

Substituting Eq. (A.4) into Eq. (A.1), we obtain:

$$R^{ij} = -\frac{8\pi G}{c^4} \cdot M^{ij} + \frac{1}{2} \cdot g^{ij} R = -\frac{8\pi G}{c^4} \cdot \left(M^{ij} - \frac{1}{2} \cdot g^{ij} M \right) . \tag{A.5}$$

Using antisymmetric tensor f_{ij} which denotes the magnitude of electromagnetic field, the electromagnetic energy tensor M^{ij} is represented as follows;

$$M^{ij} = -\frac{1}{\mu_0} \cdot \left(f^{ip} f^j_p - \frac{1}{4} \cdot g^{ij} f^{\alpha\beta} f_{\alpha\beta} \right), \quad f^{ip} = g^{i\alpha} g^{p\beta} f_{\alpha\beta} . \tag{A.6}$$

Therefore, for M, we have

$$M = M^i_i = g_{ij} M^{ij} = -\frac{1}{\mu_0} \cdot \left(g_{ij} f^{ip} f^j_p - \frac{1}{4} \cdot g_{ij} g^{ij} f^{\alpha\beta} f_{\alpha\beta} \right)$$

$$= -\frac{1}{\mu_0} \cdot \left(f^{ip} f_{ip} - \frac{1}{4} \cdot 4 f^{\alpha\beta} f_{\alpha\beta} \right) = -\frac{1}{\mu_0} \cdot \left(f^{ip} f_{ip} - f^{ip} f_{ip} \right) = 0 \tag{A.7}$$

Accordingly, substituting $M = 0$ into Eq. (A.5), we get

$$R^{ij} = -\frac{8\pi G}{c^4} \cdot M^{ij} .$$ (A.8)

Although Ricci tensor R^{ij} has 10 independent components, the major

component is the case of $i = j = 0$, i.e., R^{00}. Therefore, Eq. (A.8) becomes

$$R^{00} = -\frac{8\pi G}{c^4} \cdot M^{00} .$$ (A.9)

On the other hand, 6 components of antisymmetric tensor $f_{ij} = -f_{ji}$ are

given by electric field E and magnetic field B from the relation to Maxwell's
field equations

$$f_{10} = -f_{01} = \frac{1}{c} \cdot E_x, f_{20} = -f_{02} = \frac{1}{c} \cdot E_y, f_{30} = -f_{03} = \frac{1}{c} E_z$$
$$f_{12} = -f_{21} = B_z, f_{23} = -f_{32} = B_x, f_{31} = -f_{13} = B_y$$ (A.10)
$$f_{00} = f_{11} = f_{22} = f_{33} = 0$$

Substituting Eq. (A.10) into Eq. (A.6), we have

$$M^{00} = -\left(\frac{1}{2} \cdot \varepsilon_0 E^2 + \frac{1}{2\mu_0} \cdot B^2 \right) \approx -\frac{1}{2\mu_0} \cdot B^2 .$$ (A.11)

Finally, from Eqs. (A.9) and (A.11), we have

$$R^{00} = \frac{4\pi G}{\mu_0 c^4} \cdot B^2 = 8.2 \times 10^{-38} \cdot B^2 \quad (B \text{ in Tesla}),$$ (A.12)

where we let $\mu_0 = 4\pi \times 10^{-7}(H/m)$, $\varepsilon_0 = 1/(36\pi) \times 10^{-9}(F/m)$, $c = 3 \times 10^8 (m/s)$

$G = 6.672 \times 10^{-11}(N \cdot m^2 / kg^2)$, B is a magnetic field in Tesla and R^{00} is a major

component of spatial curvature $(1/m^2)$.

Eq. (A12) is derived from general method.

On the other hand, Levi-Civita also investigated the gravitational field
produced by a homogeneous electric or magnetic field, which was expressed
by Pauli [18]. If x^3 is taken in the direction of a magnetic field of intensity F
(Gauss unit), the square of the line element is of the form;

$$ds^2 = (dx^1)^2 + (dx^2)^2 + (dx^3)^2 + \frac{(x^1 dx^1 + x^2 dx^2)^2}{a^2 - r^2}$$
$$-\left[c_1 \exp(x^3/a) + c_2 \exp(-x^3/a)\right]^2 (dx^4)^2$$

$$(A.13)$$

where $r = \sqrt{(x^1)^2 + (x^2)^2}$, c_1 and c_2 are constants, $a = \dfrac{c^2}{\sqrt{kF}}$, k is Newtonian gravitational constant(G), and $x^1...x^4$ are Cartesian coordinates ($x^1...x^3$=space, $x^4 = ct$) with orthographic projection.

The space is cylindrically symmetric about the direction of the field, and on each plane perpendicular to the field direction the same geometry holds as in Euclidean space on a sphere of radius a, that is, the radius of curvature a is given by

$$a = \frac{c^2}{\sqrt{kF}} .$$

$$(A.14)$$

Since the relation of between magnetic field B in SI units and magnetic field F in CGS Gauss units are described as follows: $B\sqrt{\dfrac{4\pi}{\mu_0}} \Leftrightarrow F$, then the radius of curvature "a" in Eq.(A14) is expressed in SI units as the following (changing symbol, $k \to G, F \to B$):

$$a = \frac{c^2}{\sqrt{GF}} = \frac{c^2}{\sqrt{G \cdot B\sqrt{\dfrac{4\pi}{\mu_0}}}} \approx (3.484 \times 10^{18} \frac{1}{B} \quad meters) \cdot$$

$$(A.15)$$

While, scalar curvature is represented by

$$R^{00} \approx R = \frac{1}{a^2} = \frac{GB^2 \dfrac{4\pi}{\mu_0}}{c^4} = \frac{4\pi G}{\mu_0 c^4} B^2$$

$$(A.16)$$

which coincides with Eq. (A.12).

APPENDIX B: Kerr Rotating Mass Solution (derivation process of Eq. (6.13))

$$g_{00} = -\left(1 - \frac{r_g r}{r^2 + h^2 \cos^2 \theta}\right), \quad g_{33} = \frac{r^2 + h^2 \cos^2 \theta}{r^2 - r_g r + h^2},$$

where, $h = J/Mc$ ($J = angular\ momentum$), $r_g = 2GM/c^2$.

$$\alpha = \sqrt{-g_{00}} \, c^2 \Gamma_{00}^3, \quad \Gamma_{00}^3 = \frac{-g_{00,3}}{2g_{33}} \quad .$$

Calculate $\Gamma_{00}^3 = \frac{-g_{00,3}}{2g_{33}}$.

$$g_{00,3} = \frac{r_g}{r^2 + a^2 \cos^2 \theta} \cdot 1 + r_g r \cdot \frac{-2r}{\left(r^2 + a^2 \cos^2 \theta\right)^2} = \frac{r_g}{r^2 + a^2 \cos^2 \theta} - \frac{2r_g r^2}{\left(r^2 + a^2 \cos^2 \theta\right)^2}$$

$$= \frac{r_g \left(r^2 + a^2 \cos^2 \theta\right) - 2r_g r^2}{\left(r^2 + a^2 \cos^2 \theta\right)^2}$$

$$\Gamma_{00}^3 = \frac{-g_{00,3}}{2g_{33}} = \frac{2r_g r^2 - r_g \left(r^2 + h^2 \cos^2 \theta\right)}{\left(r^2 + h^2 \cos^2 \theta\right)^2} \cdot \frac{r^2 - r_g r + h^2}{2\left(r^2 + h^2 \cos^2 \theta\right)} = \frac{\left(r_g r^2 - r_g h^2 \cos^2 \theta\right) \cdot \left(r^2 - r_g r + h^2\right)}{2\left(r^2 + h^2 \cos^2 \theta\right)^3}$$

$$= \frac{r_g \left(r^2 - h^2 \cos^2 \theta\right) \cdot r^2 \left(1 - \frac{r_g}{r} + \frac{h^2}{r^2}\right)}{2\left(r^2 + h^2 \cos^2 \theta\right)^3}$$

where, Schwarzschild radius (gravitational radius): $r_g = \frac{2GM}{c^2}$ ($r_g \ll 1$),

$h^2 \ll r^2$, then $\Gamma_{00}^3 \cong \frac{r_g r^2 \left(r^2 - h^2 \cos^2 \theta\right)}{2\left(r^2 + h^2 \cos^2 \theta\right)^3}$.

$$\alpha = \sqrt{-g_{00}} \, c^2 \Gamma_{00}^3 = \sqrt{-(-1)} \cdot c^2 \cdot \frac{2GM}{c^2} \cdot \frac{1}{2} \cdot \frac{r^2 \left(r^2 - h^2 \cos^2 \theta\right)}{\left(r^2 + h^2 \cos^2 \theta\right)^3} = GM \cdot \frac{r^4 \left(1 - \frac{h^2 \cos^2 \theta}{r^2}\right)}{\left(r^2 \left(1 + \frac{h^2 \cos^2 \theta}{r^2}\right)\right)^3}$$

$$= \frac{GM}{r^2} \cdot \frac{\left(1 - \frac{h^2}{r^2} \cos^2 \theta\right)}{\left(1 + \frac{h^2}{r^2} \cos^2 \theta\right)^3} < \frac{GM}{r^2}$$

Here we used, Schwarzschild radius: $r_g = \frac{2GM}{c^2}$ ($r_g \ll 1$), $h = \frac{J}{Mc}$.

J is the angular momentum of the rotating body, M is the mass, and c is the speed of light.

APPENDIX C: Mechanical Concept of Space-time

Fundamental Concept of Space

a) Space is an infinite continuum and its structure is determined by Riemannian geometry. Space satisfies the following conditions:

b) When the infinitesimal distance regulating the distance between the two points changes by a certain physical action, the change is continuous, and the space maintains a continuum even after its change. Now, the concept of strain of continuum mechanics is very important in order to relate a spatial curvature to a practical force. Because the spatial curvature is a purely geometrical quantity. A strain field is required for the conversion of geometrical quantity to a practical force.

c) The spatial strain is defined as a localized geometrical structural change of space. It implies a change from flat space involved in zero curvature components to curved Riemann space involved in non-zero curvature components.

d) Space has the only strain-free natural state, and space always returns to the strain-free natural state, i.e., flat space, when an external physical action causing spatial strain is removed.

e) Spatial strain means some kinds of structural deformation of space, and a body filling up space is affected by the action from its spatial strain. We must distinguish space from an isolated body. An isolated body occupies an area of space by its movement. Basically, an isolated body can move in space and also can change its position.

f) In order to keep the continuity of space, the velocity of body filling up space cannot exceed the strain rate of space itself.

Since the subject of our study is a four-dimensional Riemann space as a curved space, we ascribe a great deal of importance to the curvature of space. We a priori accept that the nature of actual physical space is a four-dimensional Riemann space, that is, three dimensional space ($x=x^1$, $y=x^2$, $z=x^3$) and one dimensional time ($w=ct=x^0$), where c is the velocity of light. These four coordinate axes are denoted as x^i (i=0, 1, 2, 3).

The square of the infinitesimal distance "ds" between two infinitely proximate points x^i and x^i+dx^i is given by equation of the form:

$$ds^2 = g_{ij}dx^i dx^j,$$ (C1)

where g_{ij} is a metric tensor.

The metric tensor g_{ij} determines all the geometrical properties of space and it is a function of this space coordinate. In Riemann space, the metric tensor g_{ij} determines a Riemannian connection coefficient $\Gamma^i{}_{jk}$, and furthermore determines the Riemann curvature tensor $R^p{}_{ijk}$ or R_{pijk}, thus the geometry of space is determined by a metric tensor.

Riemannian geometry is a geometry which provides a tool to describe curved Riemann space, therefore a Riemann curvature tensor is the principal quantity. All the components of Riemann curvature tensor are zero for flat space and non-zero for curved space. If a non-zero component of Riemann curvature tensor exists, the space is not flat space, but curved space. In curved space, it is well known that the result of the parallel displacement of vector depends on the choice of the path. Further, the components of a vector differ from the initial value, after we displace a vector parallel along a closed curve until it returns to the starting point.

An external physical action such as the existence of mass energy or electromagnetic energy yields the structural deformation of space. In the deformed space region, the infinitesimal distance is given by:

$$ds'^2 = g'_{ij}dx^i dx^j,$$ (C2)

where g'_{ij} is the metric tensor of deformed space region, and we use the convected coordinates ($x'^i = x^i$).

As shown in Fig.C1, if the line element between the arbitrary two near points (A and B) in space region **S** (before structural deformation) is defined as $ds = g_i dx^i$, the infinitesimal distance between the two near points is given by

Eq. (C1): $ds^2 = g_{ij}dx^i dx^j$.

66

Let us assume that a space region **S** is structurally deformed by an external physical action and transformed to space region **T**. In the deformed space region **T**, the line element between the identical two near point (A' and B') of the identical space region newly changes, differs from the length and direction, and becomes $ds' = g'_i dx^i$.

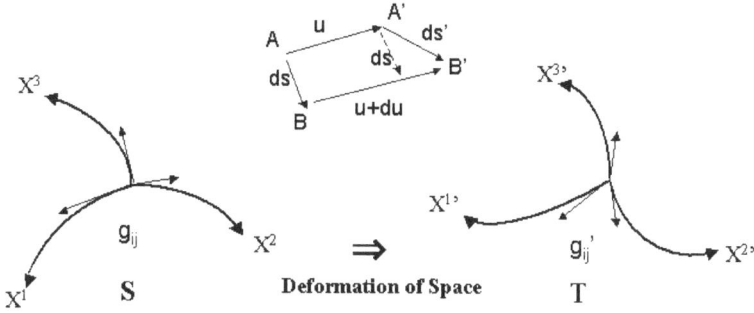

Fig. C1. Fundamental structure of Space

Therefore, the infinitesimal distance between the two near points using the convected coordinate ($x''^i = x^i$) is given by:

$$ds'^2 = g'_{ij} dx^i dx^j . \qquad (C3)$$

The g'_i is the transformed base vector from the original base vector g^i

and the g'_{ij} is the transformed metric tensor from the original metric tensor

g_{ij}. Since the degree of deformation can be expressed as the change of distance

between the two points, we get:

$$ds'^2 - ds^2 = g'_{ij} dx^i dx^j - g_{ij} dx^i dx^j = (g'_{ij} - g_{ij}) dx^i dx^j = r_{ij} dx^i dx^j . \qquad (C4)$$

Hence the degree of geometrical and structural deformation can be expressed by the quantity denoted change of metric tensor, i.e.

$$r_{ij} = g'_{ij} - g_{ij} . \qquad (C5)$$

On the other hand, the state of deformation can be also expressed by the displacement vector "u" (see Fig.C1).

From the continuum mechanics [14], the following equations are obtained:

$$du = g^i u_{i:j} dx^j ,$$ (C6)

$$ds' = ds + du = ds + g^i u_{i:j} dx^j .$$ (C7)

We use the usual notation ":" for covariant differentiation.

As is well known, the partial derivative $u_{i,j} = \dfrac{\partial u_i}{\partial x^j}$ is not tensor equation.

The covariant derivative $u_{i:j} = u_{i,j} - u_k \Gamma_{ij}^k$ is tensor equation and can be carried over into all coordinate systems.

From the usual continuum mechanics, the infinitesimal distance after deformation becomes [14]:

$$ds'^2 - ds^2 = r_{ij} dx^i dx^j = (u_{i:j} + u_{j:i} + u^k{}_{:i} u_{k:j}) dx^i dx^j .$$ (C8)

The terms of higher order than second $u^k{}_{:i} u_{k:j}$ can be neglected if the displacement is small enough value. As the actual physical space can be dealt with the minute displacement from the trial calculation of strain, we get:

$$r_{ij} = u_{i:j} + u_{j:i} .$$ (C9)

Whereas, according to the continuum mechanics [14], the strain tensor e_{ij} is given by:

$$e_{ij} = \frac{1}{2} \cdot r_{ij} = \frac{1}{2} \cdot (u_{i:j} + u_{j:i}) .$$ (C10)

So, we get:

$$ds'^2 - ds^2 = (g'_{ij} - g_{ij}) dx^i dx^j = 2e_{ij} dx^i dx^j ,$$ (C11)

where g'_{ij}, g_{ij} is a metric tensor, e_{ij} is a strain tensor, and $ds'^2 - ds^2$ is the square of the infinitesimal distance between two infinitely proximate points x^i and $x^i + dx^i$.

68

Eq. (C11) indicates that a certain geometrical structural deformation of space is shown by the concept of strain. In essence, the change of metric tensor $(g'_{ij} - g_{ij})$ due to the existence of mass energy or electromagnetic energy tensor produces the strain field e_{ij}.

Since space-time is distorted, the infinitesimal distance between two infinitely proximate points x^i and $x^i + dx^i$ is important in our understanding of the geometry of the space-time; the physical strain is generated by the difference of a geometrical metric of space-time. Namely, a certain structural deformation is described by strain tensor e_{ij}. From Eq.(C11), the strain of space is described as follows:

$$e_{ij} = 1/2 \cdot (g_{ij}' - g_{ij}). \tag{C12}$$

It is also worth noting that this result yields the principle of constancy of light velocity in Special Relativity.

Mechanics of Space

Expanding the concept of vector parallel displacement in Riemann space, the following equation has newly been obtained:

$$\omega_{\mu\nu} = R_{\mu\nu kl} dA^{kl}, \tag{C13}$$

where $\omega_{\mu\nu}$ is rotation tensor, dA^{kl} is infinitesimal areal element.

According to the nature of Riemann curvature tensor $R_{\mu\nu kl}$, $\omega_{\mu\nu}$ indicates the rotation of displacement field. Eq.(C13) indicates that a curved space produces the rotation of displacement field in the region of space. Now, the rotation tensor $\omega_{\mu\nu}$ and strain tensor e_{ij} satisfy the following differential equation in continuum mechanics:

$$\omega_{\mu\nu,j} = e_{\nu j,\mu} - e_{\mu j,\nu}. \tag{C14}$$

This equation is true on condition that the order of differential can be exchanged in a flat space. To expand above equation into a curved Riemann space, the equation shall be transformed to covariant differentiation and it is possible on condition of $\Gamma^{\alpha}_{jv} e_{\mu\alpha} = \Gamma^{\alpha}_{j\mu} e_{v\alpha}$.

Thus, we obtain:

$$\omega_{\mu v:j} = e_{vj:\mu} - e_{\mu j:v} . \tag{C15}$$

Here we use the usual notation ":" for covariant differentiation.

Eq. (C15) indicates that the displacement gradient of rotation tensor corresponds to difference of the displacement gradient of strain tensor.

Here, if we multiply both sides of Eq.(C15) by fourth order tensor denoted the nature of space $E^{ij\mu v}$ formally, we obtain:

$$E^{ij\mu v}\omega_{\mu v:j} = E^{ij\mu v}(R_{\mu vkl}dA^{kl})_{:j} = E^{ij\mu v}R_{\mu vkl:j}dA^{kl} , \tag{C16}$$

and

$$E^{ij\mu v}e_{vj:\mu} - E^{ij\mu v}e_{\mu j:v} = (E^{ij\mu v}e_{vj})_{:\mu} - (E^{ij\mu v}e_{\mu j})_{:v} = \sigma^{i\mu}_{\ :\mu} - \sigma^{iv}_{\ :v} = \Delta\sigma^{ir}_{\ :r} \tag{C17}$$

As is well known in the continuum mechanics [14], the relationship between stress tensor σ_{ij} and strain tensor e_{ml} is given by

$$\sigma^{ij} = E^{ijml}e_{ml} . \tag{C18}$$

Furthermore, the relationship between body force F^i and stress tensor σ_{ij} is given by

$$F^i = \sigma^{ij}_{\ :j} , \tag{C19}$$

from the equilibrium conditions of continuum. That is, the elastic force F^i is given by the gradient of stress tensor σ^{ij}.

Therefore, Eq.(C17) indicates the difference of body force ΔF^i. Accordingly, from Eqs (C16) and (C17), the change of body force

$$\Delta F^i (=\Delta\sigma^{ir}_{\ :r})$$

becomes

$$\Delta F^i = E^{ij\mu v}R_{\mu vkl:j}dA^{kl} . \tag{C20}$$

Here, we assume that $E^{ij\mu v}$ is constant for covariant differentiation, and A^{kl} is area element.

The stress tensor σ^{ij} is a surface force and F^i is a body force. The body force is an equivalent gravitational action because of acting all elements of space uniformly.

Eq. (C20) indicates that the gradient of Riemann curvature tensor implying space curvature produces the body force as a space strain force. The non-zero component of Eq. (C20) is just only one equation as follows:

$$F^3 = F = E^{3330}(R_{3030}A^{30})_{:3} = E^{3330} \cdot \partial(R_{3030}A^{30})/\partial r .$$
$$(C21)$$

Supplemental explanation for Derivation of Eq. (C8)

Let us consider two adjacent spatial points A and B in the unreformed space, Fig.C1, which are the end points of a line element vector ds. During the deformation, point A undergoes the displacement u and moves point A', while point B experiences a slightly different displacement $u+du$ when moving to point B'.

From Fig.C1 we read the simple vector equation

$$ds' = ds + du = ds + g^i u_{i:j}dx^i = g_i dx^i + g^i u_{i:j}dx^j \tag{1}$$

We may now write the square of the deformed line element. Since all indices are dummies, they have been chosen so that the final result looks best. When we multiply the two factors term by term and switch the notation for some dummy pairs, we obtain:

$$ds' \cdot ds' = (g_i dx^i + g^k u_{k:i}dx^i) \cdot (g_j dx^j + g^l u_{l:j}dx^j) = (g_{ij} + 2g_i \cdot g^l u_{l:j} + g^{kl}u_{k:i}u_{l:j})dx^i dx^j \tag{2}$$

From Eq. (2), we get:

$$ds'^2 - g_{ij}dx^i dx^j = (2g_i \cdot g^l u_{l:j} + g^{kl}u_{k:i}u_{l:j})dx^i dx^j \tag{3}$$

Left side of Eq. (3):

$$g'_{ij} dx^i dx^j - g_{ij}dx^i dx^j = (g'_{ij} - g_{ij})dx^i dx^j = r_{ij}dx^i dx^j \tag{4}$$

Right side of Eq. (3): considering $2g_i g^l u_{l:j} = 2\delta_i^l u_{l:j} = 2u_{i:j}$ and $g^{kl}u_{k:i}u_{l:j} = u^l{}_{:i}u_{l:j} = u^k{}_{:i}u_{k:j}$, (changes of the dummy indices l→k), then,

$$(2g_i \cdot g^l u_{l:j} + g^{kl}u_{k:i}u_{l:j})dx^i dx^j = (2u_{i:j} + u^k{}_{:i}u_{k:j}) \tag{5}$$

71

Since $r_{ij} = r_{ji}$, considering $2u_{i:j} = u_{i:j} + u_{j:i}$,

$$(g'_{ij} - g_{ij})dx^i dx^j = r_{ij}dx^i dx^j = (u_{i:j} + u_{j:i} + u^k{}_{:i}u_{k:j})dx^i dx^j$$

Finally we obtain Eq. (C8):

$$ds'^2 - ds^2 = r_{ij}dx^i dx^j = (u_{i:j} + u_{j:i} + u^k{}_{:i}u_{k:j})dx^i dx^j \qquad (6)$$

<Supplemental explanation for Derivation of Eq. (C.13)>

Let us suppose that covariant vector $A_i(P)$ at point P(x) is transported

parallel to Q(x+d1x+d2x) via path I is $A_i(Q)_{\mathrm{I}}$. On the while, covariant vector

$A_i(P)$ at point P(x) is transported parallel to Q(x+d1x+d2x) via path II is

$A_i(Q)_{\mathrm{II}}$. (Fig.C2).

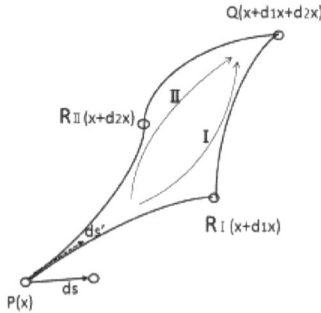

Fig. C2. Parallel transport of vector via two paths

The result of these parallel transport of vector via two paths does not differ. The difference is indicated by:

$$A_i(Q)_{\mathrm{I}} - A_i(Q)_{\mathrm{II}} = R^\rho{}_{ik\sigma}(P)A_\rho(P)d_1x^k d_2x^\sigma \qquad (1)$$

where $R^\rho{}_{ik\sigma}(p)$ is Riemann curvature tensor at point P.

Considering $A_i(Q)_I + (-A_i(Q)_{II})$, Eq. (1) indicates the quantum not returned of vector $A_i(P)$ in case that parallel transport of a vector $A_i(P)$ along a closed path has been employed at each segment of the loop $P \to R$ $I \to Q \to R\,II \to P$ and ultimately leads back to the point of departure ,i.e., original point P(x) (Fig.C3).

Mathematically, Riemann curvature tensor is the result of a difference that changed the order of the covariant derivatives as seen Eq. (2), and its non-commutative part is represented by the Riemann curvature tensor.

$$X_{i:jk} - X_{i:kj} = R^p{}_{ijk} X_p \tag{2}$$

Let us consider point R adjacent spatial point P, Fig.C3, which are the end points of a line element vector ds. If vector ds at P(x) is transported parallel to RI(x+d1x) and thence to Q(x+d1x+d2x), then parallel transport from Q(x+d1x+d2x) to original point P(x) via RII(x+d2x), the result is the new vector ds'.

Parallel transport of a vector ds along a closed path that ultimately leads back to the point of departure will result in a new vector ds' at the original point P(x); the new vector ds' differs from the original vector ds, even though the proper procedure for parallel transport has been employed at each segment of the loop. $ds'-ds$ indicates the quantum not returned of vector, also is denoted by displacement vector du. This arises from nonzero curvature of space.

Another interpretation, two adjacent spatial points P and R in the unreformed space, Fig.C3, which are the end points of a line element vector ds. During the deformation, P under goes the displacement u and moves P', while R experiences a slightly different displacement $u+du$ when moving to R'.

These phenomena are equivalent, it is not possible to identify them.

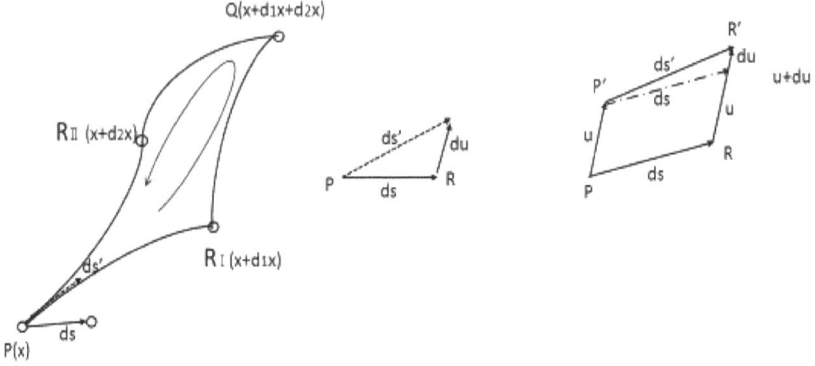

Fig. C3. Parallel transport of vector along a closed path and displacement vector

From above, we get:

$$A_i(Q)_{\mathrm{I}} - A_i(Q)_{\mathrm{II}} = ds' - ds = R^p{}_{ik\sigma}(P)A_\rho(P)d_1 x^k d_2 x^\sigma \qquad (3)$$

Since $ds'-ds = du$, infinitesimal displacement vector du is described in

$$du = du_i = u_{i;j}dx^j \qquad (4)$$

$$A_i(Q)_{\mathrm{I}} - A_i(Q)_{\mathrm{II}} = ds' - ds = du = u_{i;j}dx^j \qquad (5)$$

Apply a vector $A_\rho(P)$ in Eq. (3) to a line element vector $ds = ds_\rho$, from Eq. (5),

$$u_{i;j}dx^j = R^p{}_{ik\sigma}ds_\rho d_1 x^k d_2 x^\sigma \qquad (6)$$

Now let us multiply both sides of Eq. (6) by g^{rp}, we get following:

$$g^{rp}u_{i;j}dx^j = R^p{}_{ik\sigma}ds^r d_1 x^k d_2 x^\sigma \text{, then,}$$

$$g^{rp}u_{i;j} = R^p{}_{ik\sigma}\frac{ds^r}{dx^j}d_1 x^k d_2 x^\sigma = R^p{}_{ik\sigma}\frac{dx^r}{dx^j}d_1 x^k d_2 x^\sigma = R^p{}_{ik\sigma}\delta^r_j d_1 x^k d_2 x^\sigma \text{, finally}$$

$$g^{jp}u_{i;j} = R^p{}_{ik\sigma}d_1 x^k d_2 x^\sigma \qquad (7)$$

Multiplying both sides of Eq. (7) by g_{pm},

$$g_{pm}g^{jp}u_{i;j} = g_{pm}R^p{}_{ik\sigma}d_1 x^k d_2 x^\sigma \text{, then applying, } g_{pm}g^{jp} = \delta^j_m \text{, we get:}$$

$$u_{i;m} = R_{mik\sigma}d_1 x^k d_2 x^\sigma . \qquad (8)$$

For convenience, returns the index m to j,

$$u_{i;j} = R_{jik\sigma}d_1x^kd_2x^\sigma \qquad (9)$$

Interchanging index i, j, we get:

$$u_{j;i} = R_{ijk\sigma}d_1x^kd_2x^\sigma \qquad (10)$$

On the other hand, using the nature of the Riemann curvature tensor,

$$R_{jik\sigma} = -R_{ijk\sigma} \qquad (11)$$

Subtracting Eq. (10) from Eq. (9), we obtain:

$$u_{i;j} - u_{j;i} = -R_{ijk\sigma}d_1x^kd_2x^\sigma - R_{ijk\sigma}d_1x^kd_2x^\sigma = -2R_{ijk\sigma}d_1x^kd_2x^\sigma \qquad (12)$$

By continuum mechanics, anti-symmetric part of the displacement gradient tensor represents the rotation tensor ω_{ij},

$$\omega_{ij} = \frac{1}{2}(u_{j;i} - u_{i;j}) \qquad (13)$$

Accordingly, we obtain:

$$\omega_{ij} = \frac{1}{2}(u_{j;i} - u_{i;j}) = R_{ijk\sigma}d_1x^kd_2x^\sigma = R_{ijk\sigma}dA^{k\sigma} \qquad (14)$$

where $dA^{k\sigma}$ is the area element enclosed by the vector d_1x^k and vector d_2x^σ.

Thus, changing the dummy of incidences $i,j,k,\sigma \to \mu,\nu,k,l$, we get finally Eq.(C13):

$$\omega_{\mu\nu} = R_{\mu\nu kl}dA^{kl} \qquad (15)$$

where $\omega_{\mu\nu}$ is rotation tensor, dA^{kl} is infinitesimal areal element.

75

www.ingramcontent.com/pod-product-compliance
Lightning Source LLC
Chambersburg PA
CBHW041312210326
41599CB00003B/80